SpringerBriefs in Applied Sciences and Technology

For further volumes:
http://www.springer.com/series/8884

Anton Alexandru Kiss

Process Intensification Technologies for Biodiesel Production

Reactive Separation Processes

 Springer

Anton Alexandru Kiss
Arnhem
The Netherlands

ISSN 2191-530X   ISSN 2191-5318   (electronic)
ISBN 978-3-319-03553-6   ISBN 978-3-319-03554-3   (eBook)
DOI 10.1007/978-3-319-03554-3
Springer Cham Heidelberg New York Dordrecht London

Library of Congress Control Number: 2014930348

© The Author(s) 2014
This work is subject to copyright. All rights are reserved by the Publisher, whether the whole or part of the material is concerned, specifically the rights of translation, reprinting, reuse of illustrations, recitation, broadcasting, reproduction on microfilms or in any other physical way, and transmission or information storage and retrieval, electronic adaptation, computer software, or by similar or dissimilar methodology now known or hereafter developed. Exempted from this legal reservation are brief excerpts in connection with reviews or scholarly analysis or material supplied specifically for the purpose of being entered and executed on a computer system, for exclusive use by the purchaser of the work. Duplication of this publication or parts thereof is permitted only under the provisions of the Copyright Law of the Publisher's location, in its current version, and permission for use must always be obtained from Springer. Permissions for use may be obtained through RightsLink at the Copyright Clearance Center. Violations are liable to prosecution under the respective Copyright Law. The use of general descriptive names, registered names, trademarks, service marks, etc. in this publication does not imply, even in the absence of a specific statement, that such names are exempt from the relevant protective laws and regulations and therefore free for general use.
While the advice and information in this book are believed to be true and accurate at the date of publication, neither the authors nor the editors nor the publisher can accept any legal responsibility for any errors or omissions that may be made. The publisher makes no warranty, express or implied, with respect to the material contained herein.

Printed on acid-free paper

Springer is part of Springer Science+Business Media (www.springer.com)

Dedicated to my loving wife for her endless support and inspiration,
as well as to the precious memory of my professors and mentors

Preface

This book is among the first ones to address novel process intensification technologies, based on integrated reactive separations, for the production of biodiesel—a mixture of fatty esters. Biodiesel is a biodegradable and renewable fuel, emerging as a viable alternative to petroleum diesel. Conventional biodiesel processes still suffer from problems associated with the use of homogeneous catalysts (e.g. expensive neutralization and separation, salt waste streams) and the limitations imposed by the chemical reaction equilibrium, thus leading to severe economical and environmental penalties.

This book provides a detailed overview illustrated with several industrially relevant examples and case studies of novel reactive separation processes able to tackle the current problems and readily usable in the biodiesel production: reactive distillation, reactive absorption, reactive extraction, as well as membrane reactors and centrifugal contact separators. The integration of reaction and separation into one operating unit overcomes equilibrium limitations and provides key benefits such as reduced investment and operating costs, as well as lower plant footprint. These processes can be further enhanced by heat integration and powered by heterogeneous catalysts, to eliminate all conventional catalyst-related operations, using the raw materials and the reaction volume efficiently, while offering higher conversion and selectivity, and high energy savings compared to conventional biodiesel processes.

The focus of the book is on key aspects of these novel process intensification technologies, ranging from the working principles to conceptual design, process control, and applications. This work includes a number of novel applications relevant to industrial biodiesel and fatty esters processes, as well as results of rigorous steady-state and dynamic process simulations. The readers will have the opportunity to learn about the basic working principles, design and control of such integrated processes, while also getting a modern overview of the process intensification opportunities for biodiesel synthesis. The target audience consists primarily of students and postgraduates, chemical engineers, researchers, project leaders, technology managers, biodiesel manufacturers, and equipment suppliers.

Acknowledgments

This work is the result of many years of work in the area of reactive separation processes and process intensification technologies for biodiesel production. I am very grateful to everyone who has contributed to make it a possible success, and I especially want to thank my collaborators and co-authors of related scientific articles: Sorin Bildea and Radu Ignat (University 'Politehnica' of Bucharest, RO), Alexandre Dimian and Gadi Rothenberg (University of Amsterdam, NL), Florin Omota (Fluor, NL), and Juan Gabriel Segovia-Hernández (Universidad de Guanajuato, MX). Furthermore, I have enjoyed the interesting discussions about process intensification and reactive separations, with some remarkable persons from academia and industry, to whom I am also obliged: Žarco Olujić, Andrzej Stankiewicz and Johan Grievink (*Delft University of Technology, NL*), Sigurd Skogestad (*NTNU Trondheim, NO*), Ivar Halvorsen (*SINTEF ICT, NO*), Jan Harmsen (*Shell Global Solutions, NL*), Andrzej Górak (*Technical University of Dortmund, DE*), as well as a number of other colleagues from the chemical process industry and academia.

In addition, the excellent support and valuable help from the editors Gabriella Anderson and Anthony Doyle (*Springer, UK*) are greatly acknowledged. And last but not least, my special thanks go to my loving wife and supportive family, for their understanding, tremendous care, relentless encouragements, as well as admirable patience with me.

Contents

About the Author

Anton A. Kiss has a Ph.D. in Chemical Engineering and around 15 years of academic research and education experience, supported by a decade of industrial research experience in the area of separation technology, process intensification and process systems engineering. Currently, he works as senior project leader in the Department of Research, Development and Innovation of AkzoNobel—a Global Fortune 500 company, consistently ranked as one of the leaders in sustainability—acting as the key expert in distillation, reactive-separations, and other integrated processes. In his capacity as an award-winning researcher in separation technologies, Dr. Kiss has given many lectures at universities and conferences and has carried out more than 100 research and industrial projects. He has also supervised numerous graduation projects, and has published several textbooks and more than 50 scientific articles in peer-reviewed journals. More details are available on the personal website: http://www.tonykiss.com.

About the Author

Abstract

Process intensification technologies can drastically reduce the overall costs and improve the sustainability of many industrial processes. In particular, reactive separations offer new and exciting opportunities for manufacturing fatty acid alkyl esters involved in the industrial production of biodiesel and specialty chemicals. Biodiesel is a biodegradable and renewable fuel, emerged as a viable alternative to petroleum diesel. In spite of the recent advances, the existing biodiesel processes still suffer from serious problems associated with the use of homogeneous catalysts and the limitations imposed by the chemical reaction equilibrium, thus leading to severe economical and environmental penalties (e.g. high operating costs, salt waste streams).

This work provides a comprehensive overview illustrated with industrially relevant examples of novel reactive separation processes used in the biodiesel production: reactive distillation, reactive absorption, reactive extraction, membrane reactors or centrifugal contact separators. Remarkably, the integration of reaction and separation into one operating unit overcomes the equilibrium limitations and provides major benefits such as low investment and operating costs. Many of these reactive separation processes can be further enhanced by heat integration and also powered by heterogeneous catalysts, to eliminate all conventional catalyst-related operations, increase the efficiency in using the raw materials and the reaction volume, while offering high conversion and selectivity, and significant energy savings. The readers will learn about the working principles, design and control of integrated processes, while getting a state-of-the-art overview of the process intensification opportunities for biodiesel synthesis.

Keywords Reactive distillation · Absorption · Stripping · Extraction · Membranes · Centrifugal contactors

Chapter 1
Process Intensification Technologies

Abstract Process intensification (PI) is a process design philosophy aimed to improve process flexibility, product quality, speed to market and inherent safety, with a reduced environmental footprint. In the EU roadmap, PI is defined as a set of innovative principles applied in process and equipment design, which can bring significant benefits in terms of process and chain efficiency, lower capital and operating expenses, higher quality of products, less wastes and improved process safety. This chapter describes the concept of process intensification and the technologies based on process intensification, providing a general overview with an emphasis on the specific applications to the biodiesel production.

Process intensification (PI) is a process design philosophy that improves process flexibility, product quality, speed to market and inherent safety, with a reduced environmental footprint (Reay et al. 2013). In the European Roadmap of Process Intensification, PI is defined as a set of radically innovative principles (paradigm shift) applied in process and equipment design, which can bring significant benefits in terms of process and chain efficiency, lower capital and operating expenses, higher quality of products, less wastes and improved process safety. Essentially, process intensification refers to novel equipment, processing techniques, and process development methods that offer substantial improvements in chemical processing, and an extensive description of a PI toolbox, ordered along two dimensions: equipment and processing methods (Moulijn et al. 2008). Note that on the time-length scales, process intensification has a common area of activity with process systems engineering (PSE)—see Fig. 1.1 (Grossmann and Westerberg, 2004)—which aims to improve the decision-making for the creation and operation of the chemical supply chain, dealing with the discovery, design, manufacturing, and distribution of chemical products. The subtle relation and the friendly symbiosis between PI and PSE, has been explored in detail by Moulijn et al. (2008).

Moreover, there are several specialized books, reviews and research papers that address topics related to process intensification in the chemical industry (Stankiewicz 2003; Keil 2007; Moulijn et al. 2008; Becht et al. 2009; van Gerven and Stankiewicz 2009; Sanders et al. 2012; Nikacevic et al. 2012; Boodhoo and Harvey 2013; Reay et al. 2013), and in particular about process intensification

A. A. Kiss, *Process Intensification Technologies for Biodiesel Production*,
SpringerBriefs in Applied Sciences and Technology, DOI: 10.1007/978-3-319-03554-3_1,
© The Author(s) 2014

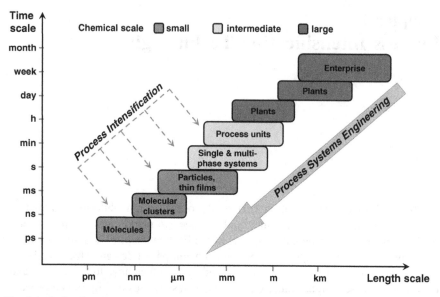

Fig. 1.1 Scale of study in process intensification versus process systems engineering

techniques applied in the biodiesel production (Qiu et al. 2010; Badday et al. 2012; Kiss and Bildea 2012; Oh et al. 2012; Ramaswamy et al. 2013).

It is worth noting that on the macro-scale of reactor and plant, the classic concept of unit operations cannot take into account the positive effect of integration. For example, in reactive separation processes the combination of reaction and separation can increase the conversion to 100 % in case of reversible reactions, by taking advantage of the *Le Chatelier* principle—pulling the equilibrium by the continuous removal of products, instead of the classic push of the equilibrium by using an excess of reactants (van Gerven and Stankiewicz 2009). Not surprisingly, Freund and Sundmacher (2008) claimed that knowledge of existing apparatuses that perform unit operations immediately narrows our creativity in search for new solutions, and they proposed to shift from unit apparatuses to functions.

A function (or a fundamental task) describes *what* should happen, and not *how* it should happen. Some examples of functions include: mass movement, chemical reaction, mixing, separation, heat transfer, phase change, temperature change, pressure change, form change, etc. An *essential function* is a function that is needed and unavoidable (e.g. chemical reaction, when converting *A* to *B*). Notably, the functional based design of chemical processes consists of fewer steps, allows the combination of more functions into one piece of equipment, and provides more freedom of equipment design (Sundmacher and Kienle 2003, 2005; Harmsen and Powell 2010; Jonker and Harmsen 2012). Due to the innovative synergistic features, the scale-up of such integrated systems also plays an important role (Harmsen 2013).

The main principles of process intensification were recently described in the research paper of van Gerven and Stankiewicz (2009), as follows:

1. *Maximize the effectiveness of intra- and inter-molecular events.* This principle is primarily about changing the kinetics of a process, which is actually the root of low conversions and selectivity, unwanted side-products and other issues. According to the simplest collision theory, the factors responsible for the effectiveness of a reaction event include: number and frequency of collisions, geometry of approach, mutual orientation of molecules in the moment of collisions, and their energy. Process intensification explores engineering methods to better control these factors.
2. *Give each molecule the same processing experience.* When all molecules undergo the same history, the process delivers ideally uniform products with minimum waste. The meso- and micro-mixing as well as temperature gradients play an important role here—along with the macroscopic residence time distribution, dead zones, or bypassing. For example, a plug-flow reactor (PFR) with gradientless, volumetric heating (e.g. by means of microwaves) is clearly much closer to the ideal described by this principle as compared to a continuous stirred-tank reactor (CSTR) with jacket heating.
3. *Optimize the driving forces at every scale and maximize the specific surface area to which these forces apply.* This principle is about the transport rates across interfaces. The resulting effect of the driving forces (e.g. concentration difference) needs to be maximized, and this is done by maximizing the interfacial area, to which that driving force applies. Increased transfer areas (or surface-to-volume ratios) can be obtained by moving from mm to μm scales of channel diameters. For example, a circular micro-channel of 400 μm in a micro-reactor delivers a specific area of 15 000 m^2/m^3. While impressive, this figure is lower than what is encountered in natural systems: e.g. capillary veins are 10 μm in diameter and have specific areas of ca. 400 000 m^2/m^3.
4. *Maximize the synergistic effects from partial processes.* Synergistic effects should be required and utilized, whenever possible and at all possible scales. Most commonly such utilization occurs in form of multi-functionality on the macro-scale, as for example in reactive separation units, where the reaction equilibrium is shifted by removing the products in situ, from the reaction environment.

These principles are not entirely new to chemical engineering, but in process intensification, they are seen as explicit goals aimed to be reached by intensified processes. Moreover, the PI interpretation of these principles often goes beyond the boundaries of the classical chemical engineering approach. A completely intensified process is successful in realizing all these process intensification principles, by making use of one or more fundamental approaches in four domains (van Gerven and Stankiewicz 2009; Gorak and Stankiewicz 2011):

1. Spatial domain (*structure*): structured environment (structured packing, structured reactors); well-defined geometry; creation of maximum specific surface

area at minimum energy expenses; creation of high mass and heat transfer rates; precise mathematical description; easy understanding; simple scale-up rules.

2. Thermodynamic domain (*energy*): alternative energy forms (electromagnetic fields) and transfer mechanisms of energy; manipulation of molecular orientation; activating and moving targeted molecules; selective, gradientless, and local energy supply.

3. Functional domain (*synergy*): integration of functions or steps; synergistic effects; better energy management; increase of overall efficiency; more compact equipment.

4. Temporal domain (*time*): timing of the events; introducing dynamics (pulsing); controlled energy input; influencing hydrodynamic behavior; increased energy efficiency; minimization of unwanted phenomena (such as fouling).

Moreover, it is worth noting that the process intensification technologies also adhere to the guiding principles for the conceptual design of safe chemical processes, thus providing *inherent safety* or *safety-by-design* (Moulijn et al. 2008; Harmsen and Powell 2010):

- *Avoid*: no extra chemicals, no solvent, no strip gas, no extra vessels, no extra pumps.
- *Small*: reduced hold-ups, low number of equipment units and inter-connections.
- *Control*: continuous processing, inherent process control (e.g. boiling systems).

The toolbox of process intensification technologies can be ordered along two dimensions: equipment and processing methods, as follows (Stankiewicz 2003; Moulijn et al. 2008; van Gerven and Stankiewicz 2009; Gorak and Stankiewicz 2011; Reay et al. 2013):

1. Process intensification equipment

   - *Chemical reactors*: spinning disk reactor (making use of centrifugal forces); static mixer reactor (specially configured tube bundles that enhance radial mixing across volume highly packed with heat transfer surface area); micro reactor (micro-structured or micro-channel reactors); and monolithic reactor (allowing chemical reaction engineering of segmented flow in micro-channels).
   - *Equipment for non-reactive systems*: rotating packed bed (based on the HiGee technology); centrifugal absorber; static mixer (precision engineered devices for the continuous mixing of fluids without the need of moving parts); and compact heat exchanger (characterized by large heat transfer area-to-volume ratio, high heat transfer coefficients, small flow passages, and laminar flow).

2. Process intensification methods

   - *Multi-functional reactors*: heat-integrated reactor, reactive separation processes (reactive distillation, reactive stripping, reactive absorption, reactive extraction, reactive crystallization, and membrane reactors), reactive comminution, reactive extrusion, and fuel cells. Among the reactive separation processes, the crown is still carried by the Eastman process that reportedly

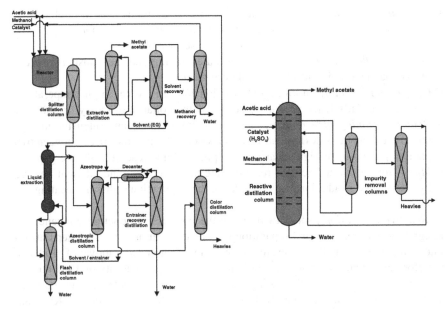

Fig. 1.2 Methyl acetate production: conventional process (*left*) versus reactive distillation (*right*)

replaced a methyl acetate production plant with a single reactive distillation column using 80 % less energy at only 20 % of the investment costs—flowsheet shown in Fig. 1.2 (Kiss 2012).

- *Hybrid separations*: dividing-wall column (combining two distillation columns into one unit); membrane distillation (separation enabled due to phase change, using a hydrophobic membrane that displays a barrier for the liquid phase, letting only the vapor phase to pass); pervaporation (separates liquid mixtures by partial vaporization through a non-porous or porous membrane); membrane adsorption; and adsorptive distillation.
- *Alternative energy sources*: solar energy, micro wave, ultrasound, electric field and centrifugal field. Using alternative energy forms aims to play a snooker game with the molecules: how to hit the right one, with right energy, at right orientation?
- *Other methods*: supercritical fluids (taking advantage that above the critical point, no distinct liquid and gas phases exist); plasma technology (ionizing the molecules of a gas thus turning it into a plasma that contains charged particles); periodic operation (such as cyclic distillation or pulsed reactors).

Remarkable, the production of biodiesel already takes advantage of using advanced process intensification technologies such as novel reactors and integrated reactive separations (Qiu et al. 2010; Kiss and Bildea 2012; Oh et al. 2012). *Novel reactors* are treated somewhere else (Qiu et al. 2010; Oh et al. 2012), and they can be summarized here as follows:

- *Static mixers*: achieve effective radial mixing as the fluids pass through it.
- *Micro-channel reactors*: improve both heat and mass transfer efficiency due to short diffusion distance and high volume to surface area.
- *Oscillatory-flow reactors*: enhance the radial mixing and the transport process by independent and controlled oscillatory motion.
- *Cavitational reactors*: allow rapid reaction rate due to the high temperature, pressure and turbulence produced locally by the collapse of cavity or bubbles.
- *Rotating/spinning tube reactors*: high mass transfer rate and intensive mixing due to the Couette flow (laminar flow of a viscous fluid in the space between two parallel plates) from high shear rate in spinning tube reactor with fine gap.
- *Microwave reactors*: increase reaction rate through direct and efficient heating by microwave irradiation, which replaces the conventional thermal heating.

Reactive separation processes improve the biodiesel production efficiency by integrating both reaction and separation (e.g. distillation or membrane separation) into a single unit that allows the simultaneous production and removal of products, therefore enhancing the reaction rate, improving the productivity and selectivity, reducing the energy use, eliminating the need for solvents, intensifying the mass and heat transfer, and ultimately leading to high-efficiency systems (Kiss and Bildea 2012). The reactive separation processes are the most promising PI technologies, as they can bring important process performance and process economic benefits. Therefore, these types of processes constitute the main topic explored in this book.

References

Badday AS, Abdullah AZ, Lee KT, Khayoon MS (2012) Intensification of biodiesel production via ultrasonic-assisted process: a critical review on fundamentals and recent development. Renew Sustain Energy Rev 16:4574–4587. doi:10.1016/j.rser.2012.04.057

Becht S, Franke R, Geißelmann A, Hahn H (2009) An industrial view of process intensification. Chem Eng Process 48:329–332. doi:10.1016/j.cep.2008.04.012

Boodhoo K, Harvey A (2013) Process intensification technologies for green chemistry: innovative engineering solutions for sustainable chemical processing. Wiley, UK

Freund H, Sundmacher K (2008) Towards a methodology for the systematic analysis and design of efficient chemical processes-Part 1: from unit operations to elementary process functions. Chem Eng Process 47:2051–2060. doi:10.1016/j.cep.2008.07.011

Gorak A, Stankiewicz A (2011) Intensified reaction and separation systems. Ann Rev Chem Biomol Eng 2:431–451. doi:10.1146/annurev-chembioeng-061010-114159

Grossmann IE, Westerberg AW (2004) Research challenges in process systems engineering. AIChE J 46:1700–1703. doi:10.1002/aic.690460902

Harmsen J, Powell JB (2010) Sustainable development in the process industries: cases and impact. Wiley-AIChE, US

Harmsen J (2013) Industrial process scale-up: a practical guide from idea to commercial implementation. Elsevier, Amsterdam

Jonker G, Harmsen J (2012) Engineering for sustainability–a practical guide for sustainable design. Elsevier, Amsterdam

Keil FJ (2007) Modeling of process intensification. Wiley-VCH, Germany

Kiss AA (2012) Applying reactive distillation. NPT Procestechnologie 19:22–24

Kiss AA, Bildea CS (2012) A review on biodiesel production by integrated reactive separation technologies. J Chem Technol Biotechnol 87:861–879. doi:10.1002/jctb.3785

Moulijn JA, Stankiewicz A, Grievink J, Gorak A (2008) Process intensification and process systems engineering: a friendly symbiosis. Comput Chem Eng 32:3–11. doi:10.1016/j.compchemeng.2007.05.014

Nikacevic NM, Huesman AEM, van den Hof PMJ, Stankiewicz AI (2012) Opportunities and challenges for process control in process intensification. Chem Eng Process 52:1–15. doi:10.1016/j.cep.2008.04.012

Oh PP, Lau HLN, Chen J, Chong MF, Choo YM (2012) A review on conventional technologies and emerging process intensification (PI) methods for biodiesel production. Renew Sustain Energy Rev 16:5131–5145. doi:10.1016/j.rser.2012.05.014

Qiu ZY, Zhao LN, Weather L (2010) Process intensification technologies in continuous biodiesel production. Chem Eng Process 49:323–330. doi:10.1016/j.cep.2010.03.005

Ramaswamy S, Huang H, Ramarao B (2013) Separation and purification technologies in biorefineries. Wiley, UK

Reay D, Ramshaw C, Harvey A (2013) Process intensification–engineering for efficiency, sustainability and flexibility, 2nd edn. Butterworth-Heinemann, UK

Stankiewicz A (2003) Reactive separations for process intensification: an industrial perspective. Chem Eng Process 42:137–144. doi:10.1016/S0255-2701(02)00084-3

Sanders JPM, Clark JH, Harmsen GJ, Heeres HJ, Heijnen JJ, Kersten SRA, van Swaaij WPM, Moulijn JA (2012) Process intensification in the future production of base chemicals from biomass. Chem Eng Process 51:117–136. doi:10.1016/j.cep.2011.08.007

Sundmacher K, Kienle A (2003) Reactive distillation: status and future directions. Wiley-VCH, Weinheim, Germany

Sundmacher K, Kienle A, Seidel-Morgenstern A (2005) Integrated chemical processes: synthesis, operation, analysis, and control. Wiley-VCH, Weinheim

van Gerven T, Stankiewicz A (2009) Structure, energy, synergy, time–the fundamentals of process intensification. Ind Eng Chem Res 48:2465–2474. doi:10.1021/ie801501y

Chapter 2
Biodiesel and Fatty Esters

Abstract This chapter provides an overview of biodiesel (basically a mixture of fatty esters) as renewable fuel, covering the market developments and trends, chemical composition and characteristics, properties and performance, complementary use as diesel fuel, main synthesis routes (e.g. esterification or transesterification), various catalysts used for manufacturing (e.g. homogeneous, solid acids and solid base catalysts) and industrial production processes (e.g. batch, continuous, supercritical, enzymatic, multi-step, reactive separations).

Fatty esters are key products of the chemical process industry, involved in various specialty chemicals with applications in the food industry, cosmetics, pharmaceuticals, plasticizers, bio-detergents and biodiesel. However, the main interest has shifted nowadays to the larger scale production of biodiesel—a mixture of fatty acid alkyl esters—hence the current strong market drive for more innovative and efficient processes (Kralova and Sjoblom 2010).

The increasing worldwide interest in biodiesel is illustrated by the tremendous increase of the production, mostly in Western Europe, North America and Asia—as shown in Fig. 2.1 (Blagoev et al. 2008; Kiss 2009). The market potential for biodiesel is actually defined and limited by the market size of the petroleum diesel. Remarkable, there is no major technical limitation on replacing fossil diesel with biodiesel, although a limitation on the feedstock—and the required arable farmland—availability does exist in practice.

The biodiesel market share suffered some changes during the recent decade, being now rather stabilized—as shown in Fig. 2.1 (bottom). An interesting development over the past years is the shift in global biodiesel market share. Europe had over 80 % capacity in 2000, but it is no longer the dominant player of the biodiesel industry, its global share accounting presently about 40 % of global capacity. Other key players emerged—such as Asia, North America, Central and South America—and they have leveled out the global biodiesel market shares. For a complete picture of the current status, Fig. 2.2 shows the biodiesel consumption worldwide and in EU (Kiss 2009). The biodiesel consumption worldwide is actually expected to grow at an average annual rate of over 5 % during 2011–2016 (Blagoev et al. 2008).

A. A. Kiss, *Process Intensification Technologies for Biodiesel Production*,
SpringerBriefs in Applied Sciences and Technology, DOI: 10.1007/978-3-319-03554-3_2,
© The Author(s) 2014

Fig. 2.1 Biodiesel
production per region and the
global biodiesel market share

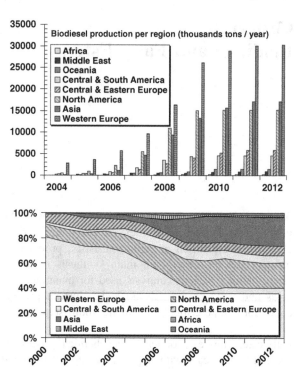

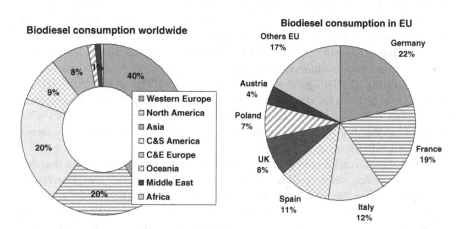

Fig. 2.2 Biodiesel consumption worldwide (*left*) and in EU (*right*)

Biodiesel is an alternative renewable and biodegradable fuel with properties
similar to petroleum diesel (Bowman et al. 2006; Balat et al. 2008; Knothe 2010).
Actually it has several advantages over petroleum diesel: it is safe, renewable,
non-toxic and biodegradable; it contains no sulfur and is a better lubricant. Despite
the chemical differences these two fuels have similar properties and performance

Table 2.1 Properties of petroleum diesel versus biodiesel

Fuel property	Diesel	Biodiesel
Fuel standard	ASTM D975	ASTM D6751
Fuel composition	C_{10}-C_{21} HC[a]	C_{12}-C_{22} FAME[a]
Kinetic viscosity, mm^2/s (at 40 °C)	1.3–4.1	1.9–6.0
Specific gravity, kg/l	0.85	0.88
Boiling point, °C	188–343	182–338
Flash point, °C	60–80	100–170
Cloud point, °C	–15 to 5	–3 to 12
Pour point, °C	–35 to –15	–15 to 10
Cetane number (ignition quality)	40–55	48–65
Stoichiometric Air/Fuel Ratio (AFR)	15	13.8
Life-cycle energy balance (energy units produced per unit energy consumed)	0.83/1	3.2/1

[a] *HC* hydrocarbons, *FAME* fatty acid methyl esters

parameters—as shown in Table 2.1 (Kiss et al. 2008). Along with its technical advantages over petroleum diesel, biodiesel brings several additional benefits to the society: rural revitalization, creation of new jobs, and less global warming.

An important characteristic of diesel fuels is the ability to auto-ignite, quantified by the cetane number (cetane index). Biodiesel not only has a higher cetane number than petroleum diesel, but also a higher flash point meaning better and safer performance. Blends of biodiesel and petroleum diesel are designated by a 'B' followed by the vol. % of biodiesel. B5 and B20—the most common blends—can be used in unmodified diesel engines. The presence of oxygen in biodiesel ($\sim$10 %) improves combustion and reduces CO, soot and hydrocarbon emissions, while slightly increasing the NOx emissions. Figure 2.3 shows the biodiesel versus petroleum diesel emissions, as well as the amount of CO_2 per distance produced by various fuels (Kiss et al. 2008). Remarkable, using B20 in trucks and buses would completely eliminate the black smoke released during acceleration and thus contribute to a cleaner air in urban areas.

2.1 Biodiesel Production Routes

Biodiesel is a mixture of fatty acid alkyl esters, produced mainly from green sources such as vegetable oils, animal fat or even waste cooking-oils from the food industry (Encinar et al. 2005; Kulkarni and Dalai 2006; Kulkarni et al. 2006; Knothe 2010; Lam et al. 2010; Maddikeri et al. 2012;). Remarkable, waste cooking-oils are much less expensive than virgin vegetable oil, and a total of over 25 million tons of waste cooking oil is generated annually, mainly in United States, China, Europe, Malaysia, Japan, Canada (Maddikeri et al. 2012). Moreover, the controversial '*food versus fuel*' competition (Knothe 2010) can be completely avoided when the raw materials used are waste vegetable oils (wvo)

Fig. 2.3 Biodiesel versus petroleum diesel emissions (*top*). Comparison of CO_2 emissions for most common fuels (*bottom*)

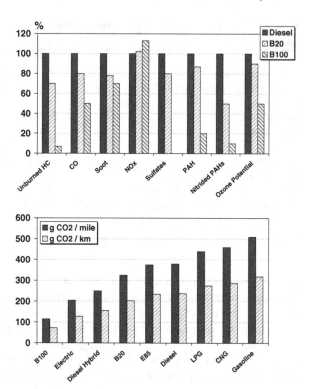

with high free fatty acids (FFA) content, non-food crops such as Jatropha (Kumar and Sharma 2005; de Oliveira et al. 2009; Kaul et al. 2010) and Mahua (Puhan et al. 2005; Kapilan and Reddy 2008; Jena et al. 2010), or even castor oil (da Silva et al. 2009; Canoira et al. 2010).

At present, employing waste and non-edible raw materials is mandatory to comply with the ecological and ethical requirements for biofuels (Feofilova et al. 2010; Nigam and Singh 2011). However, waste raw materials can contain a substantial amount of free fatty acids (Demirbas 2009; Maddikeri et al. 2012). Accordingly, the development of very efficient processes (e.g. based on reactive separation technologies) is required for the fatty esters manufacturing, in which the use of a solid catalyst is especially wanted in order to suppress the costly chemical processing steps and waste treatment (Sharma et al. 2011a, b).

As a non-petroleum-based diesel fuel, biodiesel consists of fatty acid methyl esters (FAME), currently produced by the trans-esterification of tri-alkyl glycerides (TAG) with methanol leading to glycerol by-product, or the esterification of free fatty acids (FFA) with methanol leading to water by-product. The main equilibrium reactions can be summarized as follows:

$$TAG + 3\ MeOH \leftrightarrow 3\ FAME + Glycerol\ (trans\text{-}esterification) \qquad (2.1)$$

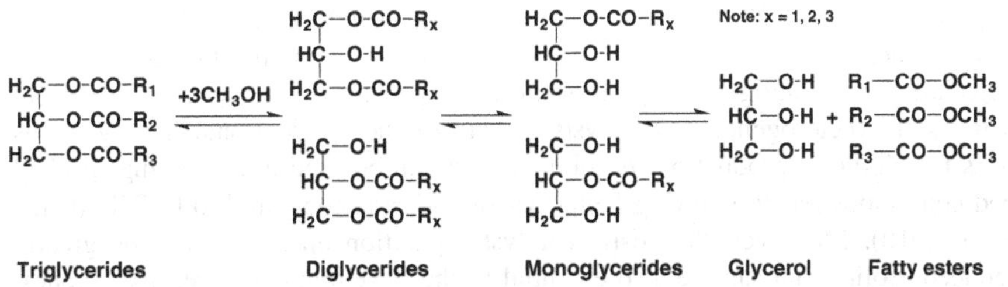

Fig. 2.4 Trans-esterification reaction of triglycerides with methanol

$$\text{FFA} + \text{MeOH} \ \leftrightarrow \ \text{FAME} + \text{H}_2\text{O} \ (\text{esterification}) \qquad (2.2)$$

The esterification reaction is typically acid catalyzed and it is rather simple. However, the trans-esterification reaction is actually more complex, proceeding with the formation of di-glycerides and mono-glycerides as intermediates, as illustrated in Fig. 2.4 (Abdullah et al. 2007). Glycerol is obtained as a by-product of the trans-esterification, typically about 10 %wt of the total amount of FAME. An excess of glycerol is therefore available on the market and it is urgent to find new convenient uses for glycerol, thus reducing also the cost of biodiesel production. Several options can be used to consume the large amount of glycerol deriving from biodiesel (Santacesaria et al. 2012), such as the use of glycerol for making commodities (e.g. glycerol hydrochlorination to chlorohydrins, or glycerol dehydration to acrolein) and for producing oxygenated additives for fuels (e.g. ethers, esters, acetals, and ketals).

In general, the trans-esterification is base catalyzed while the esterification is catalyzed by acids—although alternative acid/base catalysts could be used but at prohibitive reaction rates. The reaction time can be dramatically shortened by increasing the liquid–liquid interfacial area by various process intensification techniques (e.g. static mixers, micro-channels reactors, microwaves assisted reactors, ultrasound assisted reactors, rotating/spinning tube reactors and centrifugal contactors) or by integrating the reaction and separations steps to pull the equilibrium to full conversions (e.g. catalytic reactive distillation). After the FAME synthesis stage, there are several down-stream processing steps required for catalysts neutralization and salt removal, alcohol recovery and recycle, as well as glycerol and biodiesel purification (Hanna et al. 2005; Meher et al. 2006; Narasimharao et al. 2007; Santacesaria et al. 2012).

2.2 Catalysts for Fatty Esters Synthesis

The conventional biodiesel production is still dominated by the use of homogeneous alkaline catalysts (e.g. NaOH, KOH, K/NaOMe), leading to severe economical and environmental penalties due to the problems associated with their use

(Shahid and Jamal 2011; Atadashi et al. 2013). Development of heterogeneous catalyst such as solid and enzymes catalysts could overcome most of the problems associated with homogeneous catalysts. Presently, there is a tremendous interest in using solid (heterogeneous) catalysts instead of the conventional homogeneous ones for biodiesel production. A solid catalyst can be used in a (rotating) packed bed continuous reactors having better performance as compared to CSTRs (Chen et al. 2010). Moreover, the costly catalyst separation operations can be greatly reduced. Solid catalysts are also essential to the development of reactive-separation units for biodiesel production. Lotero et al. (2005), Helwani et al. (2009), Lee and Saka (2010), Singh and Sarma (2011), Sharma et al. (2011a, b), published recently specialized reviews about using solid catalysts in the biodiesel synthesis, while Refaat (2011) focused on using solid metal oxide catalysts such as: alkali earth metal oxides, transition metal oxides, mixed metal oxides and supported metal oxides. Clearly, these solid catalysts create new opportunities for the biodiesel production by reactive separation processes, such as catalytic reactive distillation.

The trans-esterification of tri-alkyl glycerides could be carried on in the presence of aluminosilicates as showed by Mittelbach et al. (1995) who compared the activity of layered aluminosilicates with sulfuric acid. However, the activity of the solid catalysts was lower and dependent on the operating conditions. The impregnation with sulfuric acid increased the performance of the catalysts, complete tryglicerides conversion being achieved after 4 h, at 220 °C and 52 bar. Nonetheless, leaching compromised the reusability of these catalysts. Kaita et al. (2002) used aluminium phosphate with various ratios of metal to phosphoric acid. The authors claim good activity and selectivity, while the catalyst appears to be stable. However, high temperatures (200 °C) and a large excess of methanol were still necessary.

The literature concerning the use of solid catalyst for esterification of fatty acids is much more abundant. Ion exchange resins such as Amberlyst and Nafion were proved to be effective (Chen et al. 1999; Heidekum et al. 1999), but swelling was the main problem associated with use of organic resins because it controls the accessibility of the acidic sites and therefore the reactivity. Moreover, most ion-exchange resins are not thermally-stable at high temperatures (Steinigeweg and Gmehling 2003; Kiss et al. 2006a, b), providing only limited reaction rates. Zeolites allow tailoring the catalytic properties by choosing the appropriate crystal structure, pore size, Si/Al ratio and acidic strength (Kiss et al. 2006a, b). Nevertheless, the mass transfer might impose limitations on the overall rate of the process. As a consequence, only the large-pore zeolites proved to be successful (Corma et al. 1994). Heteropoly acids supported on silica molecular sieves such as MCM41, were effective for gas-phase esterification (Verhoef et al. 1999; Mbaraka et al. 2003) under mild operating conditions (110 °C) leading to 95 % alcohol conversion. Recently, sulfated zirconia proved its activity for several acid-catalysed reactions (Yadav and Nair 1999; Omota et al. 2003a, b; Kiss et al. 2006a, b, 2008).

Table 2.2 presents an overview of the available solid acid and base catalysts for the fatty esters production by (trans-)esterification (Kiss 2010). These solid

Table 2.2 Benefits and drawbacks of acid and base catalysts tested for (trans-)esterification

Catalyst type	Benefits	Drawbacks
Ion-exchange resins (Nafion, Amberlyst)	Very high activity	Low thermal stability
	Easy regeneration	Possible leeching
TungstoPhosphoric Acid ($H_3PW_{12}O_{40}$)	Very high activity	Soluble in water
TPA-Cs salt ($Cs_{2.5}H_{0.5}PW_{12}O_{40}$)	Super acid sites	Low activity per weight
Zeolites (H-ZSM-5, Y and Beta)	Controllable acidity and hydrophobicity	Small pore size
		Low activity
Sulfated metal oxides (zirconia, titania, tin oxide)	High activity	Deactivates in water, but not in organic phase
	Thermally stable	
Niobic oxide (Nb_2O_5)	Water tolerant	Average activity
Calcium oxide/CaO	Low temperatures	Long reaction times
Calcium methoxide/Ca(OMe)$_2$	High yield, reusable	High reactants ratio
Calcium ethoxide/Ca(OEt)$_2$	High yield, short times	High reactants ratio
Li-dopped zinc oxide/ZnO	Low temperatures	Long reaction times
KF, KI, K_2CO_3	Good initial performance	Incomplete yields
KF-Eu$_2$O$_3$, KNO$_3$–Al$_2$O$_3$	Short reaction times	Subject to leeching
Single metal oxides (La$_2$O$_3$, MgO, ZnO, CeO$_2$)	High or moderate activity: $La_2O_3 > MgO \gg ZnO \sim CeO_2$	Basic strength of the active sites on surface
Mixed/supported metal oxides(Al$_2$O$_3$-SnO$_2$, Al$_2$O$_3$-ZnO, MgO-La$_2$O$_3$)	Increased surface basicity	Incomplete yields
	Resistance to FFA/moisture	
Mg–Al hydrotalcites	High activity, no leeching	Slow deactivation (fouling)
	Resistance to FFA/moisture	
Stearates of Cd, Mn and Pb	Good yield, short times	High temperatures

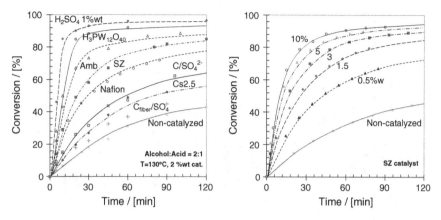

Fig. 2.5 Conversion profiles for the esterification of dodecanoic acid with 2-ethylhexanol, using various acid catalysts (*left*), and catalyzed by SZ (*right*)

catalysts are described at large in several reviews and research papers (Kiss et al. 2006a, b; Narasimharao et al. 2007; Di Serio et al. 2008; Helwani et al. 2009; Jothiramalingam and Wang 2009; Melero et al. 2009; Lee et al. 2009; Lee and Saka 2010; Refaat 2011; Semwal et al. 2011; Sharma et al. 2011a, b; Endalew et al. 2011; Singh and Sarma 2011; Santacesaria et al. 2012; Atadashi et al. 2013).

Figure 2.5 (left) provides the conversion profiles for several acid catalysts tested for the fatty esters production by esterification, such as: tungstophosphoric acid ($H_3PW_{12}O_{40}$) and its cesium salt (Cs2.5), Amberlyst-15, a styrene-based sulfonic acid, Nafion-NR50, a copolymer of tetrafluoroethene and perfluoro-2-(fluorosulfonylethoxy)-propyl vinyl ether, sulfated carbon-based catalysts (carbon fiber, mesoporous carbon), sulfated zirconia and other metal oxides (Kiss et al. 2006a, b, 2008). Sulfated zirconia (SZ) is well known for its industrial applications in a variety of processes and it can be modified with sulfate ions to form a superacidic catalyst, depending on the treatment conditions. By increasing the amount of SZ catalyst used the reaction rate, hence conversion after a certain time, can be further increased—as shown by Fig. 2.5, right (Kiss et al. 2006a, b)—making this catalyst suitable for reactive separations applications where high activity is required in a rather short time. Moreover, SZ is also very selective, thermally stable, and the re-calcination of the used catalyst can restore its original activity (Kiss et al. 2006a, b, 2008).

Although the reaction mechanism for the heterogeneous acid-catalysed esterification was shown to be in principle similar to the homogeneously catalysed one, there is an important difference that concerns the relationship between the surface hydrophobicity and the catalysts activity. This is especially true when both reactants (fatty acid and alcohol) are very lipophilic compounds. Three cases are possible, as illustrated in Fig. 2.6 (Kiss et al. 2006a, b). First, in case of one isolated Brønsted acid site surrounded by a hydrophobic environment, it is likely that the hydrophobic tail of the fatty acid would be adsorbed parallel to the

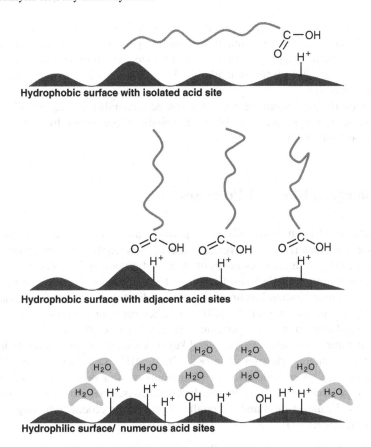

Fig. 2.6 Cartoon of the influence of the surface hydrophobicity on the catalytic activity

hydrophobic surface. Second, if there are a few acid sites in the vicinity, the fatty acid molecules could adsorb perpendicular to the surface, with the tails forming a local hydrophobic environment. Finally, in the case of a very acidic and/or hydrophilic material (many adjacent acid sites and/or hydroxy groups), the by-product water from the esterification would adsorb on the surface, and the catalyst would lose its activity since the water layer would prevent the access of fatty acid and alcohol to the catalyst (Kiss et al. 2006a, b).

Recently, Patel et al. (2013) reported the synthesis of sulfated zirconia and its characterization by various physico-chemical techniques such as energy-dispersive X-ray spectroscopy (EDS), thermal analysis using thermo-gravimetric analysis (TGA) and derivative thermo-gravimetry (DTG), Fourier transform infrared spectroscopy (FT-IR), X-ray diffraction analysis (XRD), Brunauer-Emmett-Teller (BET) surface area measurement, scanning electron microscope (SEM) and n-butyl amine acidity determination. The use of SZ catalyst was explored for the biodiesel production by esterification of oleic acid with methanol. Moreover, the influence of various reaction parameters—such as catalyst concentration, acid/ alcohol molar ratio, catalyst amount, reaction temperature and reaction time—on the catalytic performance was studied to optimize the conditions for maximum yield of 90 % methyl oleate. Remarkable, it was also possible to regenerate and reuse the catalyst.

Based on the brief literature review, it can be concluded that many solid acids and solid base catalysts are available for the biodiesel production by esterification and trans-esterification.

2.3 Industrial Biodiesel Processes

As vegetable oil is currently still too expensive, the current trend is to use less expensive raw materials. However, the use of new feedstock containing large amounts of fatty acids requires novel or improved processes for obtaining FAME through esterification of FFA and trans-esterification of tri-glycerides. At present, the most common biodiesel technologies employ homogeneous catalysts (Vicente et al. 2004; Narasimharao et al. 2007), in either batch or continuous processes where both the reaction and separation steps can create bottlenecks.

The literature overview reveals several key biodiesel processes, currently in use either at pilot and/or industrial scale (Kiss 2010, 2011; Santacesaria et al. 2012). For convenience, we provide here a brief summary of these processes, based on the catalyst type (such as homogeneous, heterogeneous, dual function, enzymatic catalysts, and catalyst-free) and the biodiesel production process type (e.g. batch, continuous, integrated):

1. *Batch processes* are conventionally used for the trans-esterification of triglyc- erides, using an acid or base catalyst. High quality virgin oil is required as raw material (<1 % FFA), otherwise a pre-treatment step is compulsory to remove the free fatty acids (Santacesaria et al. 2012). A key advantage of the batch processes is the good flexibility with respect to composition of the feedstock, but the equipment productivity is rather low and the operating costs are quite high (Hanna et al. 2005; Lotero et al. 2005). While rather simple and useful, requiring mild operating conditions (ambient pressure and temperatures lower than 100 °C), these batch processes are practically not suitable for the large- scale production of biodiesel.

2. *Continuous processes* combine both esterification and trans-esterification steps, allowing higher productivity. Nonetheless, most of these processes are still plagued by the drawbacks of using homogeneous catalysts although solid catalysts emerged in the last decade (Dale 2003; Kiss et al. 2006a, b; Yan et al. 2010; Sharma et al. 2011a, b). Several reactive distillation processes were also reported (He et al. 2005, 2006; Kiss et al. 2006a, b, 2008; Dimian et al. 2009)— as clearly illustrated in Fig. 2.7 (www.yellowdiesel.com). A commercial alternative is the ESTERFIP-H process developed by the French Institute of

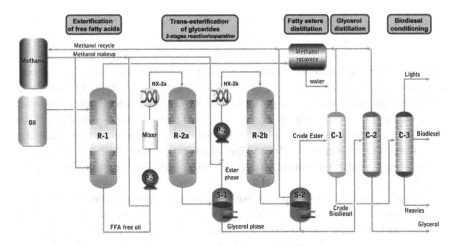

Fig. 2.7 Biodiesel production process combining esterification and trans-esterification

Petroleum (Bournay et al. 2005). This is a gas-phase process that uses a solid catalyst based on Zn and Al oxides, requiring temperatures of 210–250 °C and high pressures of 30–50 bar. As major drawbacks, the ESTERFIP-H process requires high quality refined oil and high reaction temperatures.

3. *Supercritical processes* were developed to solve the problem of oil-alcohol miscibility that hinders the kinetics of trans-esterification, as well as to take advantage of not using a catalyst at all. The operating conditions are quite severe (T > 240 °C, p > 80 bar) and therefore require special equipment (He et al. 2007; Gomez-Castro et al. 2011). However, at high temperatures, the esterification and trans-esterification reactions occur together without any problems, and the reaction rate is not affected by the presence of water by-product. The recent study of Lee and Saka (2010) emphasized the potential of non-catalytic supercritical processes and review the current status and challenging issues. Future process developments could further reduce the reaction time, by operating in the presence of a catalyst, at lower temperature and pressure—in order to render this technology more economically attractive.

4. *Enzymatic processes* have low energy requirements, as the reaction is carried out at mild conditions—ambient pressure and temperatures of 50–55 °C (Lai et al. 2005; Su et al. 2007, 2009; Chen et al. 2008; Dussan et al. 2010). However, due to the lower yields and long reaction times the enzymatic processes can not compete yet with other industrial processes (van Gerpen 2005; Demirbas 2008). In addition, research efforts are also carried out on using algae or microbial mass as raw materials (Liu and Zao 2007; Francisco et al. 2010).

5. *Multi-step processes* are somewhat simpler, as the tri-glycerides are hydrolyzed first to fatty acids that are subsequently esterified to their corresponding fatty esters in a second step (Kusdiana and Saka 2004; Minami and Saka 2006):

$$TAG + 3\ H_2O \leftrightarrow 3\ FAA + Glycerol\ (hydrolysis) \qquad (2.3)$$

$$FFA + MeOH \leftrightarrow FAME + H_2O\ (esterification) \qquad (2.4)$$

These processes are now very attractive and gain market share due to obvious advantages: high purity glycerol is obtained as by-product of the hydrolysis step, and the esterification step can be performed in conventional reactors or using solid acid catalysts in various integrated reactive-separation units (Kiss et al. 2008; Kiss 2009, 2011). Also, the use of solid catalysts avoids the neutralization and washing steps, leading to an overall simpler and more efficient process. Note that in the case when the raw materials consist mainly of FFA, only the esterification step is actually required.

6. *Reactive-separation processes* are based on esterification or trans-esterification reactions, carried out in the presence of liquid or solid catalysts, in integrated units such as: reactive distillation, reactive absorption, reactive extraction, membrane reactors or centrifugal contact separators (Kiss 2011, 2013a, b; Kiss and Bildea 2012). These integrated processes offer high conversion and selectivity, as well as increased energy efficiency. Reactive separations are the main object of this work, and therefore described in details over the next chapters.

7. *Hydro-pyrolysis processes* employ a fundamentally different chemical route as compared to the previously described manufacturing methods. Tri-glycerides are converted to fuel by hydrogenation followed by pyrolysis. The key difference is that the fuel product (second-generation biodiesel) is a mixture of long-chain hydrocarbons instead of the conventional fatty esters (Snare et al. 2009). Considering this difference of chemical composition, the correct name for this fuel product should be green-diesel and not biodiesel. The process is known as NExBTL (biomass to liquid) and it was invented by the Finnish company Neste Oil (Maki-Arvela et al. 2008; Snare et al. 2009). While it has clear advantages, this process requires more complex equipment and implies the availability of a low-cost hydrogen source.

References

Abdullah AZ, Razali N, Mootabadi H, Salamatinia B (2007) Critical technical areas for future improvement in biodiesel technologies. Environ Res Lett 2:034001. doi:10.1088/1748-9326/2/3/034001

Atadashi IM, Aroua MK, Aziz ARA, Sulaiman NMN (2013) The effects of catalysts in biodiesel production: a review. J Ind Eng Chem 19:14–26. doi:10.1016/j.jiec.2012.07.009

Balat M, Balat H (2008) A critical review of bio-diesel as a vehicular fuel. Energy Convers Manag 49:2727–2741. doi:10.1016/j.enconman.2008.03.016

Blagoev M, Bizzari S, Gubler R, Funada C, Yi Z (2008) Biodiesel. CEH Marketing Research Report 205.0000 A. Chemical Economics Handbook—SRI Consulting

Bournay L, Casanave D, Delfort B, Hillion G, Chodorge JA (2005) New heterogeneous process for biodiesel production: a way to improve the quality and the value of the crude glycerin produced by biodiesel plants. Catal Today 106:190–192. doi:10.1016/j.cattod.2005.07.181

Bowman M, Hilligoss D, Rasmussen S, Thomas R (2006) Biodiesel: a renewable and biodegradable fuel. Hydrocarbon Process 85:103–106

Canoira L, Galean JG, Alcantara R, Lapuerta M, Garcia-Contreras R (2010) Fatty acid methyl esters (FAMEs) from castor oil: production process assessment and synergistic effects in its properties. Renew Energy 35:208–217. doi:10.1016/j.renene.2009.05.006

Chen YH, Huang YH, Lin RH, Shang NC (2010) A continuous-flow biodiesel production process using a rotating packed bed. Bioresour Technol 101:668–673. doi:10.1016/j.biortech.2009.08. 081

Chen X, Du W, Liu DH, Ding FX (2008) Lipase-mediated methanolysis of soybean oils for biodiesel production. J Chem Technol Biotechnol 83:71–76. doi:10.1002/jctb.1786

Chen X, Xu Z, Okuhara T (1999) Liquid phase esterification of acrylic acid with 1-butanol catalyzed by solid acid catalysts. Appl Catal A Gen 180:261–269

Corma A, Rodriguez M, Sanchez N, Aracil J (1994) Process for the selective production of monoesters of diols and triols using zeolitic catalysts, WO9413617

da Silva ND, Batistella CB, Maciel R, Maciel MRW (2009) Biodiesel Production from castor oil: optimization of alkaline ethanolysis. Energy Fuels 23:5636–5642. doi:10.1021/ef900403j

Dale B (2003) 'Greening' the chemical industry: research and development priorities for biobased industrial products. J Chem Technol Biotechnol 78:1093–1103. doi:10.1002/jctb. 850

de Oliveira JS, Leite PM, de Souza LM, Mello VM, Silva EC, Rubim JC, Meneghetti SMP, Suarez PAZ (2009) Characteristics and composition of Jatropha Gossypifolia and Jatropha Curcas L. oils and application for biodiesel production. Biomass Bioenergy 33:449–453

Demirbas A (2008) Comparison of transesterification methods for production of biodiesel from vegetable oils and fats. Energy Convers Manag 49:125–130. doi:10.1016/j.enconman.2007. 05.002

Demirbas AH (2009) Inexpensive oil and fats feedstocks for production of biodiesel. Energy Edu Sci Technol Part A Energy Sci Res 23:1–13

Di Serio M, Tesser R, Pengmei L, Santacesaria E (2008) Heterogeneous catalysts for biodiesel production. Energy Fuels 22:207–217. doi:10.1021/ef700250g

Dimian AC, Bildea CS, Omota F, Kiss AA (2009) Innovative process for fatty acid esters by dual reactive distillation. Comput Chem Eng 33:743–750. doi:10.1016/j.compchemeng.2008.09. 020

Dussan KJ, Cardona CA, Giraldo OH, Gutierrez LF, Perez VH (2010) Analysis of a reactive extraction process for biodiesel production using a lipase immobilized on magnetic nanostructures. Bioresour Technol 101:9542–9549. doi:10.1016/j.biortech.2010.07.044

Encinar JM, Gonzalez JF, Rodriguez-Reinares A (2005) Biodiesel from used frying oil. Variables affecting the yields and characteristics of the biodiesel. Ind Eng Chem Res 44:5491–5499. doi:10.1021/ie040214f

Endalew AK, Kiros Y, Zanzi R (2011) Inorganic heterogeneous catalysts for biodiesel production from vegetable oils. Biomass Bioenergy 35:3787–3809. doi:10.1016/j.biombioe.2011.06.011

Feofilova EP, Sergeeva YE, Ivashechkin AA (2010) Biodiesel-fuel: content, production, producers, contemporary biotechnology (Review). Appl Biochem Microbiol 46:369–378. doi:10.1134/S0003683810040010

Francisco EC, Neves DB, Jacob-Lopes E, Franco TT (2010) Microalgae as feedstock for biodiesel production: carbon dioxide sequestration, lipid production and biofuel quality. J Chem Technol Biotechnol 85:395–403. doi:10.1002/jctb.2338

Gomez-Castro FI, Rico-Ramirez V, Segovia-Hernandez JG, Hernandez-Castro S (2011) Esterification of fatty acids in a thermally coupled reactive distillation column by the two-step supercritical methanol method. Chem Eng Res Des 89:480–490. doi:10.1016/j.cherd. 2010.08.009

Hanna MA, Isom L, Campbell J (2005) Biodiesel: current perspectives and future. J Sci Ind Res 64:854–857

He BB, Singh AP, Thompson JC (2005) Experimental optimization of a continuous-flow reactive distillation reactor for biodiesel production. Trans ASAE 48:2237–2243

He BB, Singh AP, Thompson JC (2006) A novel continuous-flow reactor using reactive distillation for biodiesel production. Trans ASABE 49:107–112

He H, Wang T, Zhu S (2007) Continuous production of biodiesel fuel from vegetable oil using supercritical methanol process. Fuel 86:442–447. doi:10.1016/j.fuel.2006.07.035

Heidekum A, Harmer MA, Hoelderich WF (1999) Addition of carboxylic acids to cyclic olefins catalyzed by strong acidic ion-exchange resins. J Catal 181:217–222. doi:10.1006/jcat 1998. 2300

Helwani Z, Othman MR, Aziz N, Kim J, Fernando WJN (2009) Solid heterogeneous catalysts for transesterification of triglycerides with methanol: a review. Appl Catal A 363:1–10. doi:10. 1016/j.apcata.2009.05.021

Jena PC, Raheman H, Kumar GVP, Machavaram R (2010) Biodiesel production from mixture of mahua and simarouba oils with high free fatty acids. Biomass Bioenergy 34:1108–1116. doi:10.1016/j.biombioe.2010.02.019

Jothiramalingam R, Wang MK (2009) Review of recent developments in solid acid base, and enzyme catalysts (heterogeneous) for biodiesel production via transesterification. Ind Eng Chem Res 48:6162–6172. doi:10.1021/ie801872t

Kaita J, Mimura T, Fukuda N, Hatori Y (2002) Catalysts for transesterification. US Patent 6407269

Kapilan N, Reddy RP (2008) Evaluation of methyl esters of mahua oil (Madhuca indica) as diesel fuel. J Am Oil Chem Soc 85:185–188. doi:10.1007/s11746-007-1179-5

Kaul S, Porwal J, Garg MO (2010) Parametric study of Jatropha seeds for biodiesel production by reactive extraction. J Am Oil Chem Soc 87:903–908. doi:10.1007/s11746-010-1566-1

Kiss AA (2009) Novel process for biodiesel by reactive absorption. Sep Purif Technol 69:280–287. doi:10.1016/j.seppur.2009.08.004

Kiss AA (2010) Separative reactors for integrated production of bioethanol and biodiesel. Comput Chem Eng 34:812–820. doi:10.1016/j.compchemeng.2009.09.005

Kiss AA (2011) Heat-integrated reactive distillation process for synthesis of fatty esters. Fuel Process Technol 92:1288–1296. doi:10.1016/j.fuproc.2011.02.003

Kiss AA (2013) Reactive distillation technology. In: Boodhoo K, Harvey A (eds) Process intensification technologies for green chemistry: Innovative engineering solutions for sustainable chemical processing. Wiley, New York, pp 251–274

Kiss AA (2013b) Novel applications of dividing-wall column technology to biofuel production processes. J Chem Technol Biotechnol 88:1387–1404. doi:10.1002/jctb.4108

Kiss AA, Bildea CS (2012) A review on biodiesel production by integrated reactive separation technologies. J Chem Technol Biotechnol 87:861–879. doi:10.1002/jctb.3785

Kiss AA, Dimian AC, Rothenberg G (2006a) Solid acid catalysts for biodiesel production— towards sustainable energy. Adv Synth Catal 348:75–81. doi:10.1002/adsc.200505160

Kiss AA, Dimian AC, Rothenberg G (2008) Biodiesel by catalytic reactive distillation powered by metal oxides. Energy Fuels 22:598–604. doi:10.1021/ef700265y

Kiss AA, Omota F, Dimian AC, Rothenberg G (2006b) The heterogeneous advantage: biodiesel by catalytic reactive distillation. Top Catal 40:141–150. doi:10.1007/s11244-006-0116-4

Knothe G (2010) Biodiesel: current trends and properties. Top Catal 53:714–720. doi:10.1007/ s11244-010-9457-0

Kralova I, Sjoblom J (2010) Biofuels-renewable energy sources: a review. J Dispersion Sci Technol 31:409–425. doi:10.1080/01932690903119674

Kulkarni MG, Dalai AK (2006) Waste cooking oil-an economical source for biodiesel: a review. Ind Eng Chem Res 45:2901–2913. doi:10.1021/ie0510526

Kulkarni MG, Dalai AK, Bakhshi NN (2006) Utilization of green seed canola oil for biodiesel production. J Chem Technol Biotechnol 81:1886–1893. doi:10.1002/jctb.1621

Kumar N, Sharma PB (2005) Jatropha curcus—a sustainable source for production of biodiesel. J Sci Ind Res 64:883–889

Kusdiana D, Saka S (2004) Two-step preparation for catalyst-free biodiesel fuel production— hydrolysis and methyl esterification. Appl Biochem Biotechnol 113:781–791. doi:10.1385/ ABAB:115:1-3:0781

Lai CC, Zullaikah S, Vali SR, Ju YH (2005) Lipase-catalyzed production of biodiesel from rice bran oil. J Chem Technol Biotechnol 80:331–337. doi:10.1002/jctb.1208

Lam MK, Lee MT, Mohamed AR (2010) Homogeneous, heterogeneous and enzymatic catalysis for transesterification of high free fatty acid oil (waste cooking oil) to biodiesel: a review. Biotechnol Adv 28:500–518. doi:10.1016/j.biotechadv.2010.03.002

Lee J, Saka S (2010) Biodiesel production by heterogeneous catalysts and supercritical technologies. Bioresour Technol 101:7191–7200. doi:10.1016/j.biortech.2010.04.071

Lee DW, Park YM, Lee KY (2009) Heterogeneous base catalysts for transesterification in biodiesel synthesis. Catal Surv Asia 13:63–77. doi:10.1007/s10563-009-9068-6

Liu B, Zhao Z (2007) Biodiesel production by direct methanolysis of oleaginous microbial biomass. J Chem Technol Biotechnol 82:775–780. doi:10.1002/jctb.1744

Lotero E, Liu YJ, Lopez DE, Suwannakarn K, Bruce D, Goodwin J (2005) Synthesis of biodiesel via acid catalysis. Ind Eng Chem Res 44:5353–5363. doi:10.1021/ie049157g

Maddikeri GL, Pandit AB, Gogate PR (2012) Intensification approaches for biodiesel synthesis from waste cooking oil: a review. Ind Eng Chem Res 51:14610–14628. doi:10.1021/ie301675j

Maki-Arvela P, Snare M, Eranen K, Myllyoja J, Murzin DY (2008) Continuous decarboxylation of lauric acid over Pd/C catalyst. Fuel 87:3543–3549. doi:10.1016/j.fuel.2008.07.004

Mbaraka IK, Radu DR, Lin VSY, Shanks BH (2003) Organosulfonic acid-functionalized mesoporous silicas for the esterification of fatty acid. J Catal 219:329–336. doi:10.1016/S0021-9517(03)00193-3

Meher LC, Vidya Sagar D, Naik S (2006) Technical aspects of biodiesel production by transesterification – A review. Renew Sustaine Energy Rev 10:248–268. doi:10.1016/j.rser.2004.09.002

Melero JA, Iglesias J, Morales G (2009) Heterogeneous acid catalysts for biodiesel production: current status and future challenges. Green Chem 11:1285–1308. doi:10.1039/B902086A

Minami E, Saka S (2006) Kinetics of hydrolysis and methyl esterification for biodiesel production in two-step supercritical methanol process. Fuel 85:2479–2483. doi:10.1016/j.fuel.2006.04.017

Mittlebach M, Silberholtz A, Koncar M (1995) Novel aspects concerning acid-catalyzed alcoholysis of triglycerides, In: Proceedings of the 21st world congress of the international society for fats research, The Hague, pp 497–499

Narasimharao K, Lee A, Wilson K (2007) Catalysts in production of biodiesel: a review. J Biobased Mater Bioenergy 1:19–30. doi:10.1166/jbmb.2007.0

Nigam PS, Singh A (2011) Production of liquid biofuels from renewable resources. Prog Energy Combust Sci 37:52–68. doi:10.1016/j.pecs.2010.01.003

Omota F, Dimian AC, Bliek A (2003a) Fatty acid esterification by reactive distillation. Part 1: equilibrium-based design. Chem Eng Sci 58:3159–3174. doi:10.1016/S0009-2509(03)00165-9

Omota F, Dimian AC, Bliek A (2003b) Fatty acid esterification by reactive distillation. Part 2: kinetics-based design for sulphated zirconia catalysts. Chem Eng Sci 58:3175–3185. doi:10.1016/S0009-2509(03)00154-4

Patel A, Brahmkhatri V, Singh N (2013) Biodiesel production by esterification of free fatty acid over sulfated zirconia. Renew Energy 51:227–233. doi:10.1016/j.renene.2012.09.040

Puhan S, Vedaraman N, Rambrahamam BV, Nagarajan G (2005) Mahua (Madhuca indica) seed oil: a source of renewable energy in India. J Sci Ind Res 64:890–896

Refaat AA (2011) Biodiesel production using solid metal oxide catalysts. Int J Environ Sci Technol 8:203–221

Santacesaria E, Vicente GM, Di Serio M, Tesser R (2012) Main technologies in biodiesel production: state of the art and future challenges. Catal Today 195:2–13. doi:10.1016/j.cattod.2012.04.057

Semwal S, Arora AK, Badoni RP, Tuli DK (2011) Biodiesel production using heterogeneous catalysts. Bioresour Technol 102:2151–2161. doi:10.1016/j.biortech.2010.10.080

Singh Chouhan AP, Sarma AK (2011) Modern heterogeneous catalysts for biodiesel production: a comprehensive review. Renew Sustain Energy Rev 15:4378–4399. doi:10.1016/j.rser.2011.07.112

Shahid EM, Jamal Y (2011) Production of biodiesel: a technical review. Renew Sustain Energy Rev 15:4732–4745. doi:10.1016/j.rser.2011.07.079

Sharma YC, Singh B, Korstad J (2011a) Advancements in solid acid catalysts for ecofriendly and economically viable synthesis of biodiesel. Biofuels Bioprod Biorefin 5:69–92. doi:10.1002/bbb.253

Sharma YC, Singh B, Korstad J (2011b) Latest developments on application of heterogenous basic catalysts for an efficient and eco friendly synthesis of biodiesel: a review. Fuel 90:1309–1324. doi:10.1016/j.fuel.2010.10.015

Snare M, Maki-Arvela P, Simakova IL, Myllyoja J, Murzin DY (2009) Overview of catalytic methods for production of next generation biodiesel from natural oils and fats. Russian J Phys Chem B 3:1035–1043. doi:10.1134/S1990793109070021

Steinigeweg S, Gmehling J (2003) Esterification of a fatty acid by reactive distillation. Ind Eng Chem Res 42:3612–3619. doi:10.1021/ie020925i

Su EZ, Xu WQ, Gao KL, Zheng Y, Wei DZ (2007) Lipase-catalyzed in situ reactive extraction of oilseeds with short-chained alkyl acetates for fatty acid esters production. J Mol Catal B-Enzymatic 48:28–32. doi:10.1016/j.molcatb.2007.06.003

Su EZ, You PY, Wei DZ (2009) In situ lipase-catalyzed reactive extraction of oilseeds with short-chained dialkyl carbonates for biodiesel production. Bioresour Technol 100:5813–5817. doi:10.1016/j.biortech.2009.06.077

van Gerpen J (2005) Biodiesel processing and production. Fuel Process Technol 86:1097–1107. doi:10.1016/j.fuproc.2004.11.005

Verhoef MJ, Kooyman PJ, Peters JA, van Bekkum HA (1999) A study on the stability of MCM-41-supported heteropoly acids under liquid- and gas-phase esterification conditions. Microporous Mesoporous Mater 27:365–371. doi:10.1016/S1387-1811(98)00269-8

Vicente G, Martinez M, Aracil J (2004) Integrated biodiesel production: a comparison of different homogeneous catalysts systems. Bioresour Technol 92:297–305. doi:10.1016/j.biortech.2003.08.014

Yadav GD, Nair JJ (1999) Sulfated zirconia and its modified versions as promising catalysts for industrial processes. Microporous Mesoporous Mater 33:1–48. doi:10.1016/S1387-1811(99)00147-X

Yan SL, DiMaggio C, Mohan S, Kim M, Salley SO, Ng KYS (2010) Advancements in heterogeneous catalysis for biodiesel synthesis. Top Catal 53:721–736. doi:10.1007/s11244-010-9460-5

Chapter 3
Reactive Separation Processes

Abstract This chapter explains the concept, types and key advantages of reactive separations in comparison to the conventional approach based on unit operations. Reactive separations are integrated operations combining reaction and separation into a single unit that allows the simultaneous production and removal of products. This improves productivity and selectivity, reduces the energy use, eliminates the need for solvents and leads to high-efficiency systems with green engineering attributes. In general, some of these benefits are realized by using reaction to improve separation while others are realized by using separation to improve reactions—the maximum effect being achieved when both aspects are important.

Most chemical processes involve reaction and separation operations that are typically carried out in different sections of the plant and use different equipment types—such as continuous stirred-tank reactor (CSTR), plug-flow reactor (PFR), batch reactor, distillation column—operated under a wide variety of conditions (Luyben and Yu 2008). The recent economic and environmental considerations have encouraged the chemical industry to focus on technologies based on process intensification (Harvey et al. 2003; Chen et al. 2010; Qiu et al. 2010).

The basic philosophy of the PI methods is to perform the tasks in such a manner that their combination leads to better overall performance. Since any chemical process involves unit operations for reaction and separation, most task combinations fall under the umbrella of *reactive separation processes* (Kulprathipanja 2001). Reactive separations are integrated operations that combine conveniently the reaction and separation into a single unit that allows the simultaneous production and removal of products—thus improving the productivity and selectivity, reducing the energy use, eliminating the need for solvents and leading to intensified, high-efficiency systems with green engineering attributes (Malone et al. 2003). Some of these benefits are realized by using reaction to improve separation (e.g. overcoming azeotropes, reacting away contaminants) while others are realized by using separation to improve reactions (e.g. enhancing overall rates, overcoming equilibrium limitations, improving selectivity)—the maximum effect being achieved when both aspects are important (Malone et al. 2003; Harmsen 2007, 2010). The fusion of

A. A. Kiss, *Process Intensification Technologies for Biodiesel Production*,
SpringerBriefs in Applied Sciences and Technology, DOI: 10.1007/978-3-319-03554-3_3,
© The Author(s) 2014

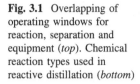

Fig. 3.1 Overlapping of operating windows for reaction, separation and equipment (*top*). Chemical reaction types used in reactive distillation (*bottom*)

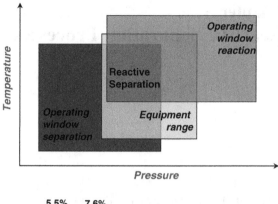

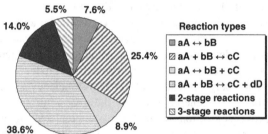

reaction and separation as one combined operation is very much appreciated for its simplicity and novelty (Kiss 2013a, b, c).

As both operations occur simultaneously in the same unit, there must be a proper match between the temperatures and pressures required for reaction and separation (Noeres et al. 2003)—as clearly illustrated by Fig. 3.1 (Kiss and Bildea 2012; Kiss 2012, 2013a, b, c). If there is no significant overlapping of the operating conditions of reaction and separation, then the combination of reaction and separation is not possible. One must also consider that working in the limited overlapping window of operating conditions is not always the optimal solution, but merely a trade-off solution (Luyben and Yu 2008).

Reactive separations are typically applied to equilibrium reactions, such as esterification, trans-esterification, etherification, hydrolysis and alkylation. Remarkable, over 1100 articles and 800 US patents on reactive distillation alone were published during the past 40 years, covering in total over 235 reaction systems (Luyben and Yu 2008). Figure 3.1 provides a convenient overview of these systems classified into various reaction types (Luyben and Yu 2008; Kiss and Bildea 2012). Moreover, Fig. 3.2 illustrates the types of reactive separations, based on the possible phase combinations (Gorak and Stankiewicz 2011)—most of them being also applicable to the biodiesel production. For more detailed information about reactive separations, the reader is kindly directed to several books and reviews published lately covering a large range of key topics: process synthesis, conceptual design, control and optimization, operation and industrial applications

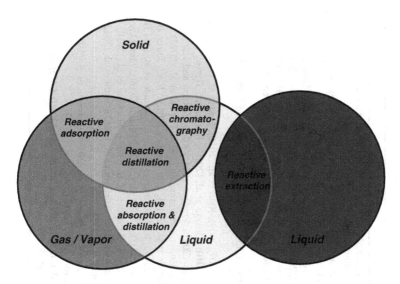

Fig. 3.2 Types of reactive separations, based on possible phase combinations

(Stankiewicz 2003; Schoenmakers and Bessling 2003; Noeres et al. 2003; Sundmacher and Kienle 2003; Sundmacher et al. 2005; Luyben and Yu 2008; Harmsen 2007, 2010; Gorak and Stankiewicz 2011).

The biodiesel production based on esterification or trans-esterification reactions can be carried out in the presence of liquid or solid catalysts, in integrated units such as: reactive distillation, reactive absorption, reactive extraction, membrane reactors or centrifugal contact separators. Table 3.1 provides a convenient overview of these reactive separations for biodiesel production (Kiss and Bildea 2012). Remarkable, reactive distillation was used for both esterification and trans-esterification reactions catalyzed by either solid or liquid catalysts. However, reactive absorption was used only for esterification catalyzed by solid acids, while reactive extraction was applied only to in situ extraction and trans-esterification of seed oils using acid/base homogeneous catalysts and various solvents. In addition, the applications of membrane reactors and pervaporation include both trans-esterification and esterification reactions. More recently, centrifugal contact separators were used for the trans-esterification of virgin oils.

The benefits of using reactive separations can be expressed among others in the energy usage. Figure 3.3 (Kiss and Bildea 2012) shows a comparison of the energy requirements for a conventional two-step process based on pre-treatment of free fatty acids (FAA) and trans-esterification of glycerides—using acid and base catalysis, respectively (Vlad et al. 2010)—versus recently reported reactive separation processes (Kiss et al. 2008; Dimian et al. 2009; Kiss 2009; 2011; Kiss and Bildea 2011, 2012). The figures are worth noting, especially considering the ongoing quest on reducing the energy use in the biodiesel production (Janulis 2004; Kiss and Bildea 2012). The specific energy use in reactive separation processes is

Table 3.1 Overview of reactive separation process for biodiesel production

Process	Reaction type	Catalyst/solvent type	References
Reactive distillation	Trans-esterification	Homogeneous (NaOH, KOH)	He et al. (2005, 2006), Prasertsit et al. (2013), Eleftheriades and von Blottnitz (2013)
		Heterogeneous (sodium ethoxide, tungstated zirconia, heteropolyacid)	Da Silva et al. (2010), Suwannakarn et al. (2009), Noshadi et al. (2011)
	Esterification	Homogeneous (H_2SO_4) Heterogeneous (ion exchange resins, mixed metal oxides, sulfated zirconia)	Cossio-Vargas et al. (2011), Omota et al. (2003), Dimian et al. (2004, 2009), Kiss et al. (2006, 2008, 2012), Kiss (2011), Ignat and Kiss (2013), Gomez-Castro et al. (2010), Hernandez et al. (2010), de Jong et al. (2010), Nguyen and Demirel (2011), Machado et al. (2011, 2013)
Reactive absorption	Esterification	Heterogeneous (mixed metal oxides, or sulfated zirconia)	Kiss (2009), Bildea and Kiss (2011), Kiss and Bildea (2011)
Reactive extraction	Trans-esterification	Homogeneous (H_2SO_4, NaOH with/out supported lipase) Solvents used: methyl or ethyl acetate, DMC, DEC, n-hexane, or no solvent at all (super-critical conditions)	Su et al. (2007, 2009), Kaul et al. (2010), Lim et al. (2010), Shuit et al. (2010), Dussan et al. (2010), Kasim and Harvey (2011), Gu et al. (2011), Pradhan et al. (2012), Zakaria and Harvey (2012), Bollin and Viamajala (2012), Porwal et al. (2012), Madankar et al. (2013), Lim and Lee (2013a, b, c), Jairurob et al. (2013a, b), Sulaiman et al. (2013), Jurado et al. (2013)
Membrane reactors	Trans-esterification	Homogeneous + Membrane (carbon, ceramic, zeolite)	Dube et al. (2007), Cao et al. (2007, 2008a, b, 2009), Tremblay et al. (2008), Baroutian et al. (2011), Machsun et al. (2010), Kapil et al. (2010), Atadashi et al. (2012), Shuit et al. (2012), Chong et al. (2013)
	Esterification	Hetero/Homogeneous + Membrane (PVA/PES)	Inoue et al. (2007), Sarkar et al. (2010), Okamoto et al. (1994), Figueiredo et al. (2010), Shi et al. (2013)
Centrifugal contact separators	Trans-esterification	Homogeneous (NaOMe, NaOEt)	Kraai et al. (2008, 2009), McFarlane et al. (2010), Oh et al. (2012), Abduh et al. (2013)

Fig. 3.3 Enegry requirements for a conventional two-step biodiesel process (*top*) versus reactive separation processes (*bottom*)

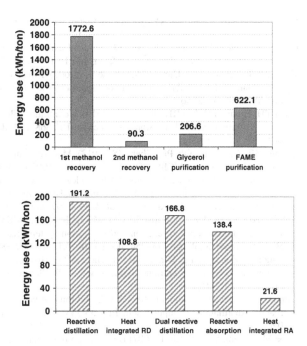

significantly lower than the FAME purification step alone in the conventional process. On top of the energy savings, the reactive separations processes also benefit from lower investment costs and reduced plant footprint—due to the less equipment used.

References

Abduh MY, van Ulden W, Kalpoe V, van de Bovenkamp HH, Manurung R, Heeres HJ (2013) Biodiesel synthesis from Jatropha curcas L. oil and ethanol in a continuous centrifugal contactor separator. Eur J Lipid Sci Tech 115:123–131. doi:10.1002/ejlt.201200173

Atadashi IM, Aroua MK, Aziz ARA, Sulaiman NMN (2012) High quality biodiesel obtained through membrane technology. J Membr Sci 421:154–164. doi:10.1016/j.memsci.2012.07.006

Baroutian S, Aroua MK, Raman AAA, Sulaiman NMN (2011) A packed bed membrane reactor for production of biodiesel using activated carbon supported catalyst. Bioresour Technol 102:1095–1102. doi:10.1016/j.biortech.2010.08.076

Bildea CS, Kiss AA (2011) Dynamics and control of a biodiesel process by reactive absorption. Chem Eng Res Des 89:187–196. doi:10.1016/j.cherd.2010.05.007

Bollin PM, Viamajala S (2012) Reactive extraction of triglycerides as fatty acid methyl esters using Lewis acidic chloroaluminate ionic liquids. Energy Fuels 26:6411–6418. doi:10.1021/ef301101d

Cao P, Dube MA, Tremblay AY (2008a) High-purity fatty acid methyl ester production from canola, soybean, palm, and yellow grease lipids by means of a membrane reactor. Biomass Bioenergy 32:1028–1036. doi:10.1016/j.biombioe.2008.01.020

Cao P, Dube MA, Tremblay AY (2008b) Methanol recycling in the production of biodiesel in a membrane reactor. Fuel 87:825–833. doi:10.1016/j.fuel.2007.05.048

Cao P, Dube MA, Tremblay AY (2009) Kinetics of canola oil transesterification in a membrane reactor. Ind Eng Chem Res 48:2533–2541. doi:10.1021/ie8009796

Cao AP, Tremblay AY, Dube MA, Morse K (2007) Effect of membrane pore size on the performance of a membrane reactor for biodiesel production. Ind Eng Chem Res 46:52–58. doi:10.1021/ie060555o

Chen YH, Huang YH, Lin RH, Shang NC (2010) A continuous-flow biodiesel production process using a rotating packed bed. Bioresour Technol 101:668–673. doi:10.1016/j.biortech.2009.08. 081

Chong MF, Chen JH, Oh PP, Chen ZS (2013) Modeling analysis of membrane reactor for biodiesel production. AIChE J 59:258–271. doi:10.1002/aic.13809

Cossio-Vargas E, Hernandez S, Segovia-Hernandez JG, Cano-Rodriguez MI (2011) Simulation study of the production of biodiesel using feedstock mixtures of fatty acids in complex reactive distillation columns. Energy 36:6289–6297. doi:10.1016/j.energy.2011.10.005

da Silva ND, Santander CM, Batistella CM, Maciel R, Maciel MRW (2010) Biodiesel production from integration between reaction and separation system: reactive distillation process. Appl Biochem Biotechnol 161:245–254. doi:10.1007/s12010-009-8882-7

de Jong MC, Zondervan E, Dimian AC, de Haan AB (2010) Entrainer selection for the synthesis of fatty acid esters by entrainer-based reactive distillation. Chem Eng Res Des 88:34–44. doi:10.1021/ie100937p

Dimian AC, Bildea CS, Omota F, Kiss AA (2009) Innovative process for fatty acid esters by dual reactive distillation. Comput Chem Eng 33:743–750. doi:10.1016/j.compchemeng.2008.09. 020

Dube MA, Tremblay AY, Liu J (2007) Biodiesel production using a membrane reactor. Bioresour Technol 98:639–647. doi:10.1016/j.biortech.2006.02.019

Dussan KJ, Cardona CA, Giraldo OH, Gutierrez LF, Perez VH (2010) Analysis of a reactive extraction process for biodiesel production using a lipase immobilized on magnetic nanostructures. Bioresour Technol 101:9542–9549. doi:10.1016/j.biortech.2010.07.044

Eleftheriades NM, von Blottnitz H (2013) Thermodynamic and kinetic considerations for biodiesel production by reactive distillation. Environ Prog Sustain Energy 32:373-376. doi:10. 1002/ep.10621

Figueiredo KCS, Salim VMM, Borges CP (2010) Ethyl oleate production by means of pervaporation-assisted esterification using heterogeneous catalysis. Braz J Chem Eng 27:609–617. doi:10.1590/S0104-66322010000400013

Gomez-Castro FI, Rico-Ramirez V, Segovia-Hernandez JG, Hernandez S (2010) Feasibility study of a thermally coupled reactive distillation process for biodiesel production. Chem Eng Process 49:262–269. doi:10.1016/j.cep.2010.02.002

Gorak A, Stankiewicz A (2011) Intensified reaction and separation systems. Ann Rev Chem Biomol Eng 2:431–451. doi:10.1146/annurev-chembioeng-061010-114159

Gu HQ, Jiang YJ, Zhou LY, Gao J (2011) Reactive extraction and in situ self-catalyzed methanolysis of germinated oilseed for biodiesel production. Energy Environ Sci 4:1337–1344. doi:10.1039/C0EE00350F

Harmsen GJ (2007) Reactive distillation: the front-runner of industrial process intensification - a full review of commercial applications, research, scale-up, design and operation. Chem Eng Process 46:774–780. doi:10.1016/j.cep.2007.06.005

Harmsen J (2010) Process intensification in the petrochemicals industry: drivers and hurdles for commercial implementation. Chem Eng Process 49:70–73. doi:10.1016/j.cep.2009.11.009

Harvey AP, Mackley MR, Seliger T (2003) Process intensification of biodiesel production using a continuous oscillatory flow reactor. J Chem Technol Biotechnol 78:338–341. doi:10.1002/ jctb.782

He BB, Singh AP, Thompson JC (2005) Experimental optimization of a continuous-flow reactive distillation reactor for biodiesel production. Trans ASAE 48:2237–2243

He BB, Singh AP, Thompson JC (2006) A novel continuous-flow reactor using reactive distillation for biodiesel production. Trans ASABE 49:107–112

Hernandez S, Segovia-Hernandez JG, Juarez-Trujillo L, Estrada-Pacheco EJ, Maya-Yescas R (2010) Design study of the control of a reactive thermally coupled distillation sequence for the esterification of fatty organic acids. Chem Eng Commun 198:1–18. doi:10.1080/00986445. 2010.493102

Ignat RM, Kiss AA (2013) Optimal design, dynamics and control of a reactive DWC for biodiesel production. Chem Eng Res Des 91:1760–1767. doi:10.1016/j.cherd.2013.02.009

Inoue T, Nagase T, Hasegawa Y, Kiyozumi Y, Sato K, Nishioka M, Hamakawa S, Mizukami F (2007) Stoichiometric ester condensation reaction processes by pervaporative water removal via acidtolerant zeolite membranes. Ind Eng Chem Res 46:3743–3750. doi:10.1021/ ie0615178

Jairurob P, Phalakornkule C, Na-udom A, Petiraksakul A (2013a) Reactive extraction of after-stripping sterilized palm fruit to biodiesel. Fuel 107:282–289. doi:10.1016/j.fuel.2013.01.051

Jairurob P, Phalakornkule C, Petiraksakul A (2013b) Single effects of reaction parameters in reactive extraction of palm fruit for biodiesel production. Chiang Mai J Sci 40:401–407

Janulis P (2004) Reduction of energy consumption in biodiesel fuel life cycle. Renewable Energy 29:861–871. doi:10.1016/j.renene.2003.10.004

Jurado MBG, Plesu V, Ruiz JB, Ruiz AEB, Tuluc A, Llacuna JL (2013) Simulation of a hybridreactive extraction unit: biodiesel synthesis. Chem Eng Trans 33:205–210. doi:10.3303/ CET1335034

Kapil A, Bhat SA, Sadhukhan J (2010) Dynamic simulation of sorption enhanced reaction processes for biodiesel production. Ind Eng Chem Res 49:2326–2335. doi:10.1021/ie901225u

Kasim FH, Harvey AP (2011) Influence of various parameters on reactive extraction of Jatropha curcas L. for biodiesel production. Chem Eng J 171:1373–1378. doi:10.1016/j.cej.2011.05. 050

Kaul S, Porwal J, Garg MO (2010) Parametric study of Jatropha seeds for biodiesel production by reactive extraction. J Am Oil Chem Soc 87:903–908. doi:10.1007/s11746-010-1566-1

Kiss AA (2009) Novel process for biodiesel by reactive absorption. Sep Purif Technol 69:280–287. doi:10.1016/j.seppur.2009.08.004

Kiss AA (2011) Heat-integrated reactive distillation process for synthesis of fatty esters. Fuel Process Technol 92:1288–1296. doi:10.1016/j.fuproc.2011.02.003

Kiss AA (2012) Applying reactive distillation. NPT Procestechnologie 19(1):22–24

Kiss AA (2013a) Advanced distillation technologies – design, control and applications. Wiley, UK

Kiss AA (2013b) Reactive distillation technology. In: Boodhoo K, Harvey A (eds) Process intensification technologies for green chemistry: innovative engineering solutions for sustainable chemical processing. Wiley, New York, pp 251–274

Kiss AA (2013c) Novel applications of dividing-wall column technology to biofuel production processes. J Chem Technol Biotechnol 88:1387–1404. doi:10.1002/jctb.4108

Kiss AA, Bildea CS (2011) Integrated reactive absorption process for synthesis of fatty esters. Bioresour Technol 102:490–498. doi:10.1016/j.biortech.2010.08.066

Kiss AA, Bildea CS (2012) A review on biodiesel production by integrated reactive separation technologies. J Chem Technol Biotechnol 87:861–879. doi:10.1002/jctb.3785

Kiss AA, Dimian AC, Rothenberg G (2008) Biodiesel by catalytic reactive distillation powered by metal oxides. Energy Fuels 22:598–604. doi:10.1021/ef700265y

Kraai GN, Van Zwol F, Schuur B, Heeres HJ, De Vries JG (2008) Two-phase (bio)catalytic reactions in a table-top centrifugal contact separator. Angew Chem Int Ed 47:3905–3908. doi:10.1002/anie.200705426

Kraai GN, Schuur B, van Zwol F, van de Bovenkamp HH, Heeres HJ (2009) Novel highly integrated biodiesel production technology in a centrifugal contactor separator device. Chem Eng J 154:384–389. doi:10.1016/j.cej.2009.04.047

Kulprathipanja S (2001) Reactive separation processes. Taylor & Francis, US

Lim S, Lee KT (2013a) Process intensification for biodiesel production from Jatropha curcas L. seeds: supercritical reactive extraction process parameters study. Appl Energy 103:712–720. doi:10.1016/j.apenergy.2012.11.024

Lim S, Lee KT (2013b) Influences of different co-solvents in simultaneous supercritical extraction and trans-esterification of Jatropha curcas L. seeds for the production of biodiesel. Chem Eng J 221:436–445. doi:10.1016/j.cej.2013.02.014

Lim S, Lee KT (2013c) Optimization of supercritical methanol reactive extraction by Response Surface Methodology and product characterization from Jatropha curcas L. seeds. Bioresour Technol 142:121–130. doi:10.1016/j.biortech.2013.05.010

Lim S, Hoong SS, Teong LK, Bhatia S (2010) Supercritical fluid reactive extraction of Jatropha curcas L. seeds with methanol: a novel biodiesel production method. Bioresour Technol 101:7169–7172. doi:10.1016/j.biortech.2010.03.134

Luyben WL, Yu CC (2008) Reactive distillation design and control. Wiley-AIChE, US

Machado GD, Aranda DA, Castier M, Cabral VF, Cardozo L (2011) Computer simulation of fatty acid esterification in reactive distillation columns. Industr Eng Chem Res 50:10176–10184. doi:10.1021/ie102327y

Machado GD, Pessoa FLP, Castier M, Aranda DAG, Cabral VF, Cardozo-Filho L (2013) Biodiesel production by esterification of hydrolyzed soybean oil with ethanol in reactive distillation columns: simulation studies. Industr Eng Chem Res 52:9461–9469. doi:10.1021/ie400806q

Machsun AL, Gozan M, Nasikin M, Setyahadi S, Yoo YJ (2010) Membrane microreactor in biocatalytic transesterification of triolein for biodiesel production. Biotechnol Bioprocess Eng 15:911–916. doi:10.1007/s12257-010-0151-7

Madankar CS, Pradhan S, Naik SN (2013) Parametric study of reactive extraction of castor seed (Ricinus communis L.) for methyl ester production and its potential use as bio lubricant. Industr Crops Prod 43:283–290. doi:10.1016/j.indcrop.2012.07.010

Malone MF, Huss RS, Doherty MF (2003) Green chemical engineering aspects of reactive distillation. Environ Sci Technol 37:5325–5329. doi:10.1021/es034467w

McFarlane J, Tsouris C, Birdwell JF, Schuh DL, Jennings HL, Palmer Boitrago AM, Terpstra SM (2010) Production of biodiesel at the kinetic limit in a centrifugal reactor/separator. Industr Eng Chem Res 49:3160–3169. doi:10.1021/ie901229x

Nguyen N, Demirel Y (2011) Using thermally coupled reactive distillation columns in biodiesel production. Energy 36:4838–4847. doi: 10.1016/j.energy.2011.05.020

Noeres C, Kenig EY, Gorak A (2003) Modelling of reactive separation processes: Reactive absorption and reactive distillation. Chem Eng Process 42:157–178. doi:10.1016/S0255-2701(02)00086-7

Noshadi I, Amin NAS, Parnas RS (2011) Continuous production of biodiesel from waste cooking oil in a reactive distillation column catalyzed by solid heteropolyacid: Optimization using response surface methodology (RSM). Fuel 94:156–164. doi: 10.1016/j.fuel.2011.10.018

Oh PP, Lau HLN, Chen J, Chong MF, Choo YM (2012) A review on conventional technologies and emerging process intensification (PI) methods for biodiesel production. Renew Sustain Energy Rev 16:5131–5145. doi:10.1016/j.rser.2012.05.014

Okamoto KI, Yamamoto M, Noda S, Semoto T, Otoshi Y, Tanaka K, Kita H (1994) Vaporpermeation-aided esterification of oleic acid. Ind Eng Chem Res 33:849–853. doi: 10.1021/ie00028a010

Omota F, Dimian AC, Bliek A (2003a) Fatty acid esterification by reactive distillation. Part 1: Equilibrium-based design. Chem Eng Sci 58:3159–3174. doi: 10.1016/S0009-2509(03)00165-9

Omota F, Dimian AC, Bliek A (2003b) Fatty acid esterification by reactive distillation. Part 2: Kinetics-based design for sulphated zirconia catalysts. Chem Eng Sci 58:3175–3185. doi: 10.1016/S0009-2509(03)00154-4

Porwal J, Bangwal D, Garg MO, Kaul S, Harvey AP, Lee JGM, Kasim FH, Eterigho EJ (2012) Reactive-extraction of pongamia seeds for biodiesel production. J Sci Ind Res 71:822–828

Pradhan S, Madankar CS, Mohanty P, Naik SN (2012) Optimization of reactive extraction of castor seed to produce biodiesel using response surface methodology. Fuel 97:848–855. doi: 10.1016/j.fuel.2012.02.052

Prasertsit K, Mueanmas C, Tongurai C (2013) Transesterification of palm oil with methanol in a reactive distillation column. Chem Eng Process 70:21–26. doi: 10.1016/j.cep.2013.05.011

Qiu ZY, Zhao LN, Weather L (2010) Process intensification technologies in continuous biodiesel production. Chem Eng Process 49:323–330. doi:10.1016/j.cep.2010.03.005

Sarkar B, Sridhar S, Saravanan K, Kale V (2010) Preparation of fatty acid methyl ester through temperature gradient driven pervaporation process. Chem Eng J 162:609–615. doi: 10.1016/j.cej.2010.06.005

Schoenmakers HG, Bessling B (2003) Reactive and catalytic distillation from an industrial perspective. Chem Eng Process 42:145–155. doi:10.1016/S0255-2701(02)00085-5

Shi W, He B, Cao Y, Li J, Yan F, Cui Z, Zou Z, Guo S, Qian X (2013) Continuous esterification to produce biodiesel by SPES/PES/NWF composite catalytic membrane in flow-through membrane reactor: Experimental and kinetic studies. Bioresour Technol 129:100–107. doi: 10.1016/j.biortech.2012.10.039

Shuit SH, Lee KT, Kamaruddin AH, Yusup S (2010a) Reactive extraction and in situ esterification of Jatropha curcas L. seeds for the production of biodiesel. Fuel 89:527–530. doi: 10.1016/j.fuel.2009.07.011

Shuit SH, Lee KT, Kamaruddin AH, Yusup S (2010b) Reactive extraction of Jatropha curcas L. seed for production of biodiesel: Process optimization study. Environ Sci Technol 44:4361–4367. doi: 10.1021/es902608v

Shuit SH, Ong YT, Lee KT, Subhash B, Tan SH (2012) Membrane technology as a promising alternative in biodiesel production: A review. Biotechnol Adv 30:1364–1380. doi: 10.1016/j.biotechadv.2012.02.009

Stankiewicz A (2003) Reactive separations for process intensification: an industrial perspective. Chem Eng Process 42:137–144. doi:10.1016/S0255-2701(02)00084-3

Su EZ, Xu WQ, Gao KL, Zheng Y, Wei DZ (2007) Lipase-catalyzed in situ reactive extraction of oilseeds with short-chained alkyl acetates for fatty acid esters production. J Mol Catal B Enzym 48:28–32. doi: 10.1016/j.molcatb.2007.06.003

Su EZ, You PY, Wei DZ (2009) In situ lipase-catalyzed reactive extraction of oilseeds with shortchained dialkyl carbonates for biodiesel production. Bioresour Technol 100:5813–5817. doi: 10.1016/j.biortech.2009.06.077

Sulaiman S, Aziz ARA, Aroua MK (2013) Reactive extraction of solid coconut waste to produce biodiesel. J Taiwan Inst Chem Eng 44:233–238. doi: 10.1016/j.jtice.2012.10.008

Sundmacher K, Kienle A (2003) Reactive distillation: Status and future directions. Wiley-VCH, Weinheim

Sundmacher K, Kienle A, Seidel-Morgenstern A (2005) Integrated chemical processes: synthesis, operation, analysis, and control. Wiley-VCH, Weinheim

Suwannakarn K, Lotero E, Ngaosuwan K, Goodwin JG (2009) Simultaneous free fatty acid esterification and triglyceride transesterification using a solid acid catalyst with in situ removal of water and unreacted methanol. Ind Eng Chem Res 48:2810–2818. doi: 10.1021/ie800889w

Tremblay AY, Cao P, Dube MA (2008) Biodiesel production using ultralow catalyst concentrations. Energy Fuels 22:2748–2755. doi: 10.1021/ef700769v

Vlad E, Bildea CS, Plesu V, Marton G, Bozga G (2010) Design of biodiesel production process from rapeseed oil. Rev Chim 61:595–603. doi:10.3303/CET1021212

Zakaria R, Harvey AP (2012) Direct production of biodiesel from rapeseed by reactive extraction/in situ transesterification. Fuel Process Technol 102:53-60. doi: 10.1016/j.fuproc.2012.04.026

Chapter 4
Property Models and Process Simulation

Abstract Process simulation is used for the design, development, analysis, and optimization of any chemical process, including biodiesel production. Property models provide the basis of computer simulations, being responsible for the accuracy or inaccuracy of the simulation results. This chapter summarizes the requirements for a successful simulation of a biodiesel production process: recommended property models, best simulation approach (e.g. shortcut, rigorous or hybrid methods), use of equilibrium or rate-based models, and reaction kinetics.

The modeling and simulation of a conventional or reactive separation process for biodiesel production, requires a minimum number of properties for all the chemical species: the feed oil (typically a mixture of tri-alkyl glycerides and fatty acids), methanol or ethanol as (trans)-esterification reactants, intermediate reaction products (mono-glycerides and di-glycerides), the by-products (glycerol and water) and the main biodiesel product (a mixture of fatty acid esters). These properties include: normal boiling point, critical temperature and pressure, acentric factor, liquid density or liquid molar volume, ideal gas capacity, liquid heat capacity, heat of vaporization and vapor pressure (Yuan et al. 2005; Zong et al. 2010).

Fatty acids, methanol, ethanol, glycerol and water are common chemical species hence their properties are accessible in databases. Moreover, the thermophysical properties of fatty acid esters are available in the NIST (National Institute of Standards and Technology) ThermoData Engine (http://trc.nist.gov/tde.html). Depending on the composition of any oil/fat feedstock used in the biodiesel production process, there is a corresponding blend of fatty acid esters in the biodiesel product (Kiss 2010). Therefore, the most difficulties are typically associated with representing the composition and calculating the properties of triglycerides.

Zong et al. (2010) considered the feed oil as a mixture of mixed tri-alkyl glycerides. A mixed triglyceride is composed of a glycerol fragment and three fatty acid fragments. Experimental data was used to correlate the contribution of each fragment to the heat of vaporization. In principle, the procedure could be applied to other properties of interest to process design. However, such a detailed approach seems rather difficult and not really necessary in practice.

A. A. Kiss, *Process Intensification Technologies for Biodiesel Production*, 35
SpringerBriefs in Applied Sciences and Technology, DOI: 10.1007/978-3-319-03554-3_4,
© The Author(s) 2014

Siang et al. (2003) considered the oil as a mixture of simple triglycerides in order to estimate the critical properties required for thermodynamic calculations, while the amount of tri-glycerides in the mixture was based on the fatty acids composition. Espinosa et al. (2002) and Ndiaye et al. (2006) built a pseudo-triglyceride in which the CH_2 and double bonds ratio is similar to the fatty acid composition in the feed oil. The thermodynamic properties of this hypothetical molecule were found by group contribution methods. In another approach by Morad et al. (2000), the thermodynamic properties were found based on fatty acid composition, but the feed oil was considered as one pseudo-triglyceride.

Many methods are available that can predict the properties of the chemical species based on the given structure of the molecule (Poling et al. 2004). These can be applied to calculate the properties of the oil feedstock, although not all the methods are suitable due to the complex structure of triglycerides and their high molecular weight. As the boiling temperature, the critical parameters and the acentric factor are not experimentally available they are treated as characteristic parameters in thermodynamic models. Various vapor-liquid equilibrium (VLE) studies with different prediction methods for these parameters give similar results. The phase behavior of oil, fatty acids and other components is important when designing or simulating reactive separation processes. Still, due to the lack of experimental data, the phase behavior is predicted by various estimation methods (Poling et al. 2004; Kuramochi et al. 2009).

The recommended methods include the original UNIFAC method (UNIQUAC Functional-group Activity Coefficients) and UNIFAC-DMD (Dortmund modified) for VLE, as well as UNIFAC-LLE for modeling the liquid-liquid equilibrium. For example, Dimian et al. (2004, 2009) and Kiss et al. (2006a, b, 2008, 2012) used Aspen Plus as a commercial process simulation software for designing different reactive separation processes for biodiesel production. For the case studies presented in the next chapters, the physical properties required for the simulation and the binary interaction parameters for the methanol-water and acid-ester pairs were available in the Aspen Plus database of pure components (Aspen Technology 2010), while missing binary interaction parameters were estimated using the UNIFAC-DMD group contribution method. Notably, the physical properties of the fatty acids and methyl esters are quite similar.

Vapor pressure is one of the most important properties with a critical effect in modeling reactive separations. Figure 4.1 shows the vapor pressure of the most common fatty acids and their corresponding methyl esters (Kiss 2009). At ambient pressure the boiling points are rather high—these values being in line with reported data from literature (Yuan et al. 2005).

The residue curves map and the ternary diagram of the mixture fatty acid-methanol-water is presented in Fig. 4.2 (Kiss 2011). As phase splitting is employed to separate the acid-water mixture, the reactive separation units are modeled using VLLE data. Moreover, the possibility of phase splitting inside the reactive separation unit must be accounted for, as free water phase may deactivate the solid acid catalysts (Okuhara 2002; Kiss et al. 2006a, b, 2008).

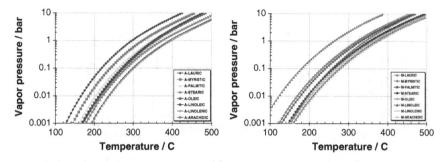

Fig. 4.1 Vapor pressure of fatty acids and their corresponding methyl esters

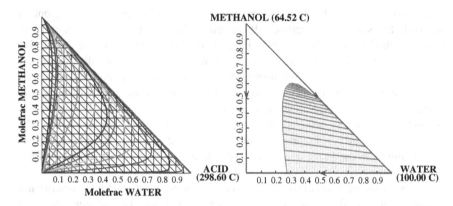

Fig. 4.2 Residue curve map and ternary diagram for the mixture fatty acid-methanol-water

The residue curve map (RCM) of simultaneous phase equilibrium and chemical equilibrium can be conveniently represented for all four component mixtures in a bi-dimensional diagram, using special coordinates, as X_1 (acid + water) and X_2 (acid + ester). Figure 4.3 left (Omota et al. 2003a, b; Kiss et al. 2006b) illustrates the esterification of lauric acid with 2-ethylhexanol. The RCM diagram shows the segregation in two liquid phases, organic and aqueous, separated by a boundary connecting the two azeotropes. There is also a third homogeneous region (only water phase) in the right corner, not visualized here because of the scale. The trajectories converge from the azeotropes of 2-ethylhexanol to the 2-ethylhexyl laurate, which is the highest boiler. The heterogeneous region can be easily avoided in practice by operating at temperatures above 100 °C. Similarly, Fig. 4.3 (right) illustrates the esterification of lauric acid with methanol (Kiss et al. 2006b). Here, the reactants are nodes, while the products are saddles. Again, there is a heterogeneous region, but no azeotrope methanol/water. The trajectories emerge from the methanol to lauric acid, passing along the ester saddle.

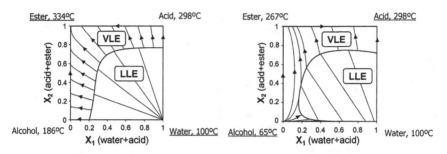

Fig. 4.3 Reactive residue curve maps for the esterification of the lauric acid with 2-ethylhexanol (*left-side*) and with methanol (*right-side*)

The simulation of a biodiesel production process can be performed using one of the available simulation methods illustrated in Table 4.1: rigorous, shortcut or hybrid method (Kiss 2010). Each of these methods has clear advantages but also specific drawbacks. Moreover, the requirements in terms of input data can differ considerably and it has a great impact as the information reflects directly in the results. Obviously, the quality of output is determined by the quality of the input, as illustrated by the well-known '*garbage in–garbage out*' concept. If the property method is improperly selected, the result of the simulation will most likely be inaccurate. Similarly, if incorrect kinetic data is used as input, or some chemicals are missing from the list of components, the output of the simulation is unlikely to be any informative.

The rigorous method is favored due to the accurate results, but it is virtually not feasible in practice due to the amount of input data required. On the other hand, the shortcut method is very handy due to its quickness in making results available, but it provides merely low-fidelity models with very limited practical applications. Therefore, for practical reasons, the hybrid approach gives the best results as it combines the advantages of both rigorous and shortcut methods, while limiting the overall disadvantages due to the synergistic effect.

The reactive separation columns from the case studies described in the next chapters, were simulated in Aspen Plus using the rigorous RADFRAC unit with equilibrium or rate-based models, and considering the three-phase balances prior to simulations (Kiss et al. 2006a, b, 2008; Dimian et al. 2009; Kiss 2009, 2010, 2011). The fatty components were conveniently lumped into one fatty acid and fatty ester, according to: R–COOH + CH$_3$OH ↔ R–COO–CH$_3$ + H$_2$O. Dodecanoic (lauric) acid/ester was selected as lumped component due to availability of experimental results and VLE parameters for this system (Kiss et al. 2006a, b, 2008; Kiss 2009; Dimian et al. 2009; Kiss and Bildea 2011). Moreover, sulfated zirconia was considered as solid acid catalyst due to the available kinetics (Kiss et al. 2008; Dimian et al. 2009). The esterification reaction is a second order reversible reaction, with the reaction rate given by:

$$r = (k_1 W_{cat}) C_{Acid} C_{Alcohol} - (k_2 W_{cat}) C_{Ester} C_{Water} \qquad (4.1)$$

Table 4.1 Simulation methods for biodiesel production: requirements, benefits and drawbacks

	Rigorous method	Shortcut method	Hybrid method
Requirements	Properties for all species	Properties for single fatty acid/ester/triglyceride	Single or reduced list of fatty acid/ester/TG
	VLL data and BIP's for all pairs of components	VLL data for the system ester/glycerol/alcohol	Short list of VLL data and BIP's for components
	Kinetic parameters for all reactions possible	Asumed conversion (no kinetic parameters)	Reduced list of kinetic parameters, few reactions
Benefits	Easy optimization of reaction and separation	Simple model Fast simulations	Optimization possible for reaction and separation
	High fidelity model Usable for many plants	Easy-to-build mass and energy balance	Certain ability to compare various feedstocks
	Easy comparison for various feedstocks	No data needed for all species present	Better model fidelity Fast simulations for RTO
Drawbacks	Slow simulations and convergence problems	No comparison possible for various feedstocks	More effort to build component list and get kinetic parameters
	Expensive measurements	Low-fidelity model	More work to find VLL data and regress BIP's
	Limited RTO and model based control usage	Less ability to use RTO	

where k_1 and k_2 are the kinetic constants for the direct (esterification) and reverse (hydrolysis) reactions, W_{cat} is the amount of catalyst, and C_n are molar concentrations of the components. As water is continuously removed from the system, the reverse hydrolysis reaction is extremely slow hence the second term of the reaction rate can be practically neglected. Note that the activation energy is $E_a = 55$ kJ/mol, and the Arrhenius factor is $A = 1.2 \cdot 10^5$ m$^3 \cdot$kmol$^{-1} \cdot$s^{-1} (for molar concentrations) or $A = 250$ kmol$\cdot$m$^3 \cdot$kg$^{-2} \cdot$s^{-1} (when mass concentrations are used).

The literature review shows that process simulation plays an important role in the biodiesel production, being used for the process design, development, analysis, as well as optimization. Several property models are available, providing the basis of computer simulations that can use shortcut, rigorous or hybrid methods. The recommended property models are mainly based on group contribution methods, such as UNIFAC. In addition, the simulations of biodiesel processes need also basic data related to the equilibrium and reaction kinetics.

References

Aspen Technology (2010) Aspen Plus: User guide, vol 1 & 2. Aspen Technology, Burlington, US
Dimian AC, Omota F, Bliek A (2004) Entrainer-enhanced reactive distillation. Chem Eng
    Process 43:411–420. doi:10.1016/S0255-2701(03)00125-9
Dimian AC, Bildea CS, Omota F, Kiss AA (2009) Innovative process for fatty acid esters by dual
    reactive distillation. Comput Chem Eng 33:743–750. doi:10.1016/j.compchemeng.2008.09.020
Espinosa S, Fornari T, Bottini SB, Brignole EA (2002) Phase equilibria in mixtures of fatty oils
    and derivatives with near critical fluids using the GC-EOS model. J Supercrit Fluids
    23:91–102. doi:10.1016/S0896-8446(02)00025-6
Kiss AA (2009) Novel process for biodiesel by reactive absorption. Sep Purif Technol
    69:280–287. doi:10.1016/j.seppur.2009.08.004
Kiss AA (2010) Separative reactors for integrated production of bioethanol and biodiesel.
    Comput Chem Eng 34:812–820. doi:10.1016/j.compchemeng.2009.09.005
Kiss AA (2011) Heat-integrated reactive distillation process for synthesis of fatty esters. Fuel
    Process Technol 92:1288–1296. doi: 10.1016/j.fuproc.2011.02.003
Kiss AA, Bildea CS (2011) Integrated reactive absorption process for synthesis of fatty esters.
    Bioresour Technol 102:490–498. doi:10.1016/j.biortech.2010.08.066
Kiss AA, Dimian AC, Rothenberg G (2006a) Solid acid catalysts for biodiesel production—
    Towards sustainable energy. Adv Synth Catal 348:75–81. doi:10.1002/adsc.200505160
Kiss AA, Omota F, Dimian AC, Rothenberg G (2006b) The heterogeneous advantage: biodiesel
    by catalytic reactive distillation. Top Catal 40:141–150. doi:10.1007/s11244-006-0116-4
Kiss AA, Dimian AC, Rothenberg G (2008) Biodiesel by catalytic reactive distillation powered
    by metal oxides. Energy Fuels 22:598–604. doi:10.1021/ef700265y
Kiss AA, Segovia-Hernandez JG, Bildea CS, Miranda-Galindo EY, Hernandez S (2012) Reactive
    DWC leading the way to FAME and fortune. Fuel 95:352–359. doi:10.1016/j.fuel.2011.12.064
Kuramochi H, Maeda K, Kato S, Osako M, Nakamura K, Sakai S (2009) Application of UNIFAC
    models for prediction of vapor-liquid and liquid-liquid equilibria relevant to separation and
    purification processes of crude biodiesel fuel. Fuel 88:1472–1477. doi:10.1016/j.fuel.2009.01.017
Morad NA, Mustafa Kamal MAA, Panau F, Yew TW (2000) Liquid specific heat capacity
    estimation for fatty acids, triacylglycerols, and vegetable oils based on their fatty acid
    composition. J Am Oil Chem Soc 77:1001–1005. doi:10.1007/s11746-000-0158-6
Ndiaye PM, Franceschi E, Oliveira D, Dariva C, Tavares FW, Oliveira JV (2006) Phase behavior
    of soybean oil, castor oil and their fatty acid ethyl esters in carbon dioxide at high pressures.
    J Supercrit Fluids 37:29–37. doi:10.1016/j.supflu.2005.08.002
Okuhara T (2002) Water-tolerant solid acid catalysts. Chem Rev 102:3641–3665. doi:10.1021/
    cr0103569
Omota F, Dimian AC, Bliek A (2003a) Fatty acid esterification by reactive distillation. Part 1:
    Equilibrium-based design. Chem Eng Sci 58:3159–3174. doi:10.1016/S0009-2509(03)
    00165-9
Omota F, Dimian AC, Bliek A (2003b) Fatty acid esterification by reactive distillation. Part 2:
    Kinetics-based design for sulphated zirconia catalysts. Chem Eng Sci 58:3175–3185. doi:10.
    1016/S0009-2509(03)00154-4
Poling BE, Prausnitz JM, O'Connell JP (2004) The properties of gases and liquids. McGraw-Hill,
    New York
Siang LC, Manan ZA, Sarmidi MR (2003) Simulation modeling of the phase behaviour of palm
    oil with supercritical carbon dioxide. In: Proceedings of international conference on chemical
    and bioprocess engineering. University Malaysia Sabah, Kota Kinabatu, pp 427–434
Yuan W, Hansen AC, Zhang Q (2005) Vapor pressure and normal boiling point predictions for
    pure methyl esters and biodiesel fuels. Fuel 84:943–950. doi:10.1016/j.fuel.2005.01.007
Zong L, Ramanathan S, Chen CC (2010) Fragment-based approach for estimating thermophys-
    ical properties of fats and vegetable oils for modeling biodiesel production processes. Ind Eng
    Chem Res 49:876–886. doi:10.1021/ie900513k

Chapter 5
Reactive Distillation Technology

Abstract Reactive distillation (RD) processes integrate reaction and distillation, the reactants being converted with the simultaneous separation of the products. This chapter describes the use of reactive distillation technology to produce biodiesel by esterification of free fatty acids, or trans-esterification of various oils (triglycerides). An overview of the reported experimental and simulation studies is provided, covering also the process design, dynamics and control, as well as more complex RD configurations based on dividing-wall column technology.

Reactive distillation (RD) is especially attractive in systems where certain chemical and phase equilibrium conditions co-exist. The reaction and distillation take place in the same zone of a distillation column, the reactants being converted with the simultaneous separation of the products and recycle of unused reactants. As the products must be separated from reactants by distillation, this implies that the products should be lighter and/or heavier than the reactants. The ideal case is when one product is the lightest and the other product is the heaviest, with the reactants being the intermediate boiling components (Luyben and Yu 2008).

RD setups may consist of multiple catalyst systems, vapor and liquid traffic over the catalyst, separation, mass flow, and enthalpy exchange—all these processes being optimally integrated in a single unit, a characteristic feature of process intensification. By continuously removing the products, RD makes it possible to use only the stoichiometric reactants ratio (neat operation) and to pull the equilibrium to high conversions. This is in contrast to the typical practice of using an excess of one of the reactants to push the equilibrium toward the desired products, at the cost of having to recover and recycle the unreacted raw material. Table 5.1 lists the most important applications of reactive distillation, reported so far: (trans-)esterification, etherification, hydrolysis, (de-)hydration, alkylation, isomerization, (de-)hydrogenation, amination, condensation, chlorination, nitration, etc.—all these reactions being limited by the chemical equilibrium (Sundmacher and Kienle 2003; Harmsen 2007; Kiss 2012, 2003a, b, c).

The production of fatty esters by reactive distillation was proposed for all sort of alcohols—ranging from methanol, ethanol, propanol to 2-ethyl hexanol (Steinigeweg and Gmehling 2003; von Scala et al. 2003; Omota et al. 2003a, b; Kiss

A. A. Kiss, *Process Intensification Technologies for Biodiesel Production*, 41
SpringerBriefs in Applied Sciences and Technology, DOI: 10.1007/978-3-319-03554-3_5,
© The Author(s) 2014

Table 5.1 Main industrial applications of reactive distillation

Reaction type	Catalyst/internals
Alkylation	
Alkyl benzene from ethylene/propylene and benzene	Zeolite β, molecular sieves
Amination	
Amines from ammonia and alcohols	H_2 and hydrogenation catalyst
Carbonylation	
Acetic acid from CO and methanol/dimethyl ether	Homogeneous
Condensation	
Diacetone alcohol from acetone	Heterogeneous
Bisphenol-A from phenol and acetone	N/A
Trioxane from formaldehyde	Strong acid catalyst, zeolite ZSM-5
Esterification	
Methyl acetate from methanol and acetic acid	H_2SO_4, Dowex 50, Amberlyst-15
Ethyl acetate from ethanol and acetic acid	N/A
2-methyl propyl acetate from 2-methyl propanol and acid	Katapak-S
Butyl acetate from butanol and acetic acid	Cation exchange resin
Fatty acid methyl esters from fatty acids and methanol	H_2SO_4, Amberlyst-15, Metal oxides
Fatty acid alkyl esters from fatty acids and alkyl alcohols	H_2SO_4, Amberlyst-15, Metal oxides
Cyclohexyl carboxylate from cyclohexene and acids	Ion exchange resin bags
Etherification	
MTBE from isobutene and methanol	Amberlyst-15
ETBE from isobutene and ethanol	Amberlyst-15/pellets, structured
TAME from isoamylene and methanol	Ion exchange resin
DIPE from isopropanol and propylene	ZSM 12, Amberlyst-36, zeolite
Hydration/Dehydration	
Mono ethylene glycol from ethylene oxide and water	Homogeneous
Hydrogenation/Dehydrogenation	
Cyclohexane from benzene	Alumina supported Ni catalyst
MIBK from benzene	Cation exchange resin with Pd/Ni
Hydrolysis	
Acetic acid and methanol from methyl acetate and water	Ion exchange resin bags
Acrylamide from acrylonitrile	Cation exchanger, copper oxide
Isomerization	
Iso-parafins from n-parafins	Chlorinated alumina and H_2
Nitration	
4-Nitrochlorobenzene from chlorobenzene and nitric acid	Azeotropic removal of water
Transesterification	
Ethyl acetate from ethanol and butyl acetate	Homogeneous
Diethyl carbonate from ethanol and dimethyl carbonate	Heterogeneous
Vinyl acetate from vinyl stearate and acetic acid	N/A
Unclassified reactions	
Monosilane from trichlorsilane	Heterogeneous
Methanol from syngas	$Cu/Zn/Al_2O_3$ and inert solvent
DEA from monoethanolamine and ethylene oxide	N/A

et al. 2006a, b, 2008; Machado et al. 2011, 2013), in a variety of configurations such as: entrainer enhanced reactive distillation (Dimian et al. 2004; de Jong et al. 2010), dual reactive distillation (Dimian et al. 2009), thermally coupled reactive distillation (Gomez-Castro et al. 2010, 2011) or reactive distillation with side-draw (Thotla and Mahajani 2009). Consequently, a number of studies were also reported on the biodiesel production by reactive distillation—either experimental work or simulation results (Kiss et al. 2006a, b, 2008; He et al. 2006; Suwannakarn et al. 2009; da Silva et al. 2010; Kiss 2011; Simasatitkul et al. 2011; Kiss and Bildea 2012).

5.1 RD Experimental Studies

He et al. (2005) reported a comprehensive study of biodiesel preparation from canola oil in a continuous-flow reactive distillation system. Optimization of process variables (feed methanol to triglycerides molar ratio, reaction time, reboiler temperature, catalyst concentration, methanol circulation mode, and catalyst formulation) was studied experimentally and analyzed statistically. An experimental design was used in the experiments, and statistical multiple response regression models were employed for process optimization. Under the operating conditions explored, product yields ranged from 41.5 to 94.9 %, productivity varied from 16 to 55.8 kmol/m^{3}h (5.6–19.5 m^3/m^{3}h), and soap formation ranged from 4.44 to 29.1 mol/100 mol (0.19–1.27 wt%). For different optimization criteria, the following optimum variable ranges were found: feed molar ratio from 3.65:1 to 4.50:1, reaction time from 3.76 to 5.56 min, reboiler temperature from 100 to 130 °C, and catalyst concentration from 0.13 to 0.24 mol/mol. Notably, the synergic effect of the process variable combinations affected the system efficiency more significantly than the individual process variables. The maximal product yields and productivity were 98.8 % and 55.6 kmol/m^{3}h (18.5 m^3/m^{3}h), respectively. However, when soap formation was minimized, the yield and productivity were only 72 % and 9.3 kmol/m^{3}h (3.1 m^3/m^{3}h), respectively. Therefore, the optimization of the RD reactor system should be based on the maximization of product yield and reactor productivity.

In a follow-up study, He et al. (2006) developed and investigated a novel reactor system using reactive distillation for biodiesel preparation from canola oil and methanol. The goal was to drastically reduce the use of excess methanol while maintaining a high methanol to glyceride molar ratio inside the RD reactor by recycling a small amount of methanol within the system. The effect of the methanol to glycerides ratio was studied on a lab-scale perforated-tray RD reactor system. Product parameters such as methyl ester content, glycerides, and methanol content were analyzed. The results showed that the RD reactor with a methanol to glycerides molar ratio of 4:1 (in which the use of methanol was cut down by 66 %) gave a satisfactory yield and conversion rate, at 65 °C. Total reaction time in the pre-reactor and RD column was about 3 min, which is 20–30 times shorter than in typical batch processes. The productivity of the RD reactor system was about 6.6 m^3 biodiesel per m^3 reactor volume per hour, which is 6–10 times higher than that of batch and existing continuous-flow processes (He et al. 2006).

Suwannakarn et al. (2009) reported the use of a three-phase reaction for the simultaneous esterification of FFA and trans-esterification of tri-alkyl glycerides using model mixtures of TAG and FFA—0–25 wt% lauric acid (HLa) in trica-prylin (TCp)—to simulate the waste grease and to permit an easy distinction between reactions. Trans/esterification with methanol was carried out using a commercial tungstated zirconia (WZ) as solid acid catalyst in a well-stirred semi-batch reactor at 130 °C and atmospheric pressure. Methanol was continuously charged to the reactor, while the unreacted methanol and water by-product were continuously removed by vaporization, thereby favoring esterification. It is worth noting that esterification occurred four times faster than trans-esterification. Under the reported reaction conditions and in the presence of WZ catalyst, water was produced not only by esterification but also by the dehydration of methanol. Due to the presence of water, the TAG conversion to fatty esters followed two routes: 1. direct TAG trans-esterification, and 2. TAG hydrolysis followed by esterification. The WZ catalyst underwent 37 % deactivation after three cycles of 2-h reaction but completely recovered its original activity after a re-calcination in flowing air.

Da Silva et al. (2010) presented an efficient process using reactive distillation columns applied to biodiesel production from soybean oil and bioethanol. The key variables affecting the biodiesel production process are: catalyst concentration, reaction temperature, level of agitation, ethanol/soybean oil molar ratio, reaction time, and raw material type. Experimental design was used to optimize the catalyst concentration (0.5–1.5 wt%) and the ethanol to soybean oil molar ratio (3:1–9:1) under a reflux rate of 83 ml/min and 6 min reaction time.

Noshadi et al. (2011) investigated the optimal conditions for obtaining fatty acid methyl esters from waste cooking oil in a reactive distillation column, using a heteropolyacid as catalyst ($H_3PW_{12}O_{40} \cdot 6H_2O$). The 80 mm column had eight trays and a kettle reboiler with electrical heating, total condenser and reflux splitter. The feed stream was introduced as liquid on the top tray. Methanol was the top product, being recycled. The bottom product, a mixture of glycerol and FAME, was separated in a decanter. The effect of four operating parameters was investigated: total feed flow, feed temperature, reboiler duty and the methanol/oil ratio. The optimum conditions were determined to be 116.23 mol/h total feed, 29.9 °C feed temperature, 1.3 kW reboiler duty, and 67.9 methanol/oil ratio. The actual FAME yield was 93.94 %.

Prasertsit et al. (2013) determined the effects of the reboiler temperature, amount of KOH catalyst, methanol to oil molar ratio and residence time on the methyl esters purity in a laboratory-scale reactive distillation packed column. The results indicated that from the empty column, the system reached the steady state in 8 h. The optimal operating condition were at a reboiler temperature 90 °C, a methanol to oil molar ratio of 4.5:1.0, KOH of 1 wt% with respect to oil, and 5 min of residence time in the RD column leading to 92.27 % purity of the methyl esters—which actually requires further purification in order to be used as fuel.

Eleftheriades and von Blottnitz (2013) built bench-scale RD unit, and interpreted the results from kinetic and thermodynamic perspectives. Analysis of samples taken from the pre-reactor and the reboiler product was performed using a conversion-viscosity chart that allowed the determination of conversion from the measured viscosity. They found that 50–60 % conversion was possible at ambient temperatures, with stoichiometric feed ratio and a stirrer. Despite the large difference in boiling points, the separation of methanol from the reactant and products mix was limited. Under all modeled scenarios reported, an appreciable amount of methanol would be lost to the product stream—concluding that the affinity of the methanol to remain in solution was greater than what was expected based on the boiling points alone.

5.2 RD Simulation Studies and Process Design

Simasatitkul et al. (2011) proposed the use of RD for biodiesel production by trans-esterification of soybean oil and methanol, catalyzed by sodium hydroxide. The simulation results showed that a suitable configuration of the RD column consists of only three reactive stages. Methanol and soybean oil should be fed into the column on the first stage. The optimal operating conditions were at the molar feed ratio of methanol and oil at 4.5:1, molar reflux ratio of 3, and reboiler duty of 1.6×10^7 kJ/h. The effect of the most important operating and design parameters on the performance of RD was also analyzed.

Machado et al. (2011) presented the computational steady-state simulations of fatty acid esters (biodiesel) production by fatty acid esterification in a RD column using niobium oxide as catalyst. Reaction rates were considered explicitly in the model of each stage. The procedures and formulation were validated by comparison of simulations results with data available in the literature. Two new cases for fatty acid esters (biodiesel) production were simulated, showing that conversions close to 99 % are possible with the proper choice of operating conditions, as demonstrated by the sensitivity analysis. The simulations results proved to be useful for the proper design of processes that use reactive distillation columns for biodiesel production.

In a follow-up study, Machado et al. (2013) presented steady-state simulations of fatty acid esters production in a reactive distillation column by esterification of anhydrous ethanol with a feedstock representative to the composition of the soybean oil (hydrolyzed soybean oil). The sensitivity analysis showed that the best operating conditions were the minimum reflux ratio of 0.001 and 15 theoretical stages. High conversion values (close to 99 %) were possible with the proper choice of these operating conditions.

Kiss et al. (2006a, b, 2008) and Kiss (2011) reported the use of solid acid catalysts (e.g. sulfated zirconia) applied in an esterification process based on catalytic reactive distillation. Such an integrated process is able to shift the chemical equilibrium to completion and preserve the solid catalyst activity by continuously removing the products, while also leading to lower investment and operating costs (Kiss et al. 2006a, b, 2008; Kiss 2010, 2011). The integrated reactive distillation process was designed according to previously reported process synthesis methods for reactive separations (Schembecker and Tlatlik 2003; Noeres et al. 2003). Rigorous simulations embedding experimental results were performed using Aspen Plus (AspenTech 2010). The RD column was simulated using the rigorous RADFRAC unit with the RateSep (rate-based) model, and explicitly considering three phase balances.

Figure 5.1 (top) presents the flowsheet of this biodiesel process based on conventional reactive distillation, as reported by Kiss et al. (2008). The reactive distillation column (RDC) consists of a core reactive zone completed by rectifying and stripping separation sections, whose extent depends on the separation behavior of the reaction mixture. Since methanol is being consumed in the reaction, a mixture of water-acid separates easily in the top. After decanting, the organic phase is returned as reflux while water is withdrawn as product. High purity FAME is obtained from the bottom stream after removing methanol in an additional flash. The reference flowsheet presented in Fig. 5.1 (top) is relatively simple, with just a few operating units, two cold streams that need to be pre-heated (fatty acid and alcohol) and two hot streams that have to be cooled down (top water and bottom fatty esters). Accordingly, the heat-integration was performed by applying heuristic rules (Hamed et al. 1996; Dimian and Bildea 2008). Consequently, a feed-effluent heat exchanger (FEHE) replaced each of the two heat exchangers HEX1 and HEX2. Figure 5.1 (btm) illustrates the improved process design including heat-integration around the RD column (Kiss 2011). The hot bottom product of the column (FAME) is used to pre-heat both reactants: the fatty acid and alcohol feed streams. Notably, there is no longer need for an external hot utility to pre-heat the reactants and no additional heat exchanger is needed by the heat-integrated setup.

The main design parameters—such as column size, catalyst loading, and feed condition—are conveniently listed in Table 5.2, while Table 5.3 provides the complete mass balance of the heat-integrated RD process (Kiss 2011). High conversion of the reactants is achieved, with the productivity of the RD unit exceeding 20 kg fatty ester/kg catalyst/h and the purity specifications over 99.9 wt% for the final biodiesel product (FAME stream).

Figure 5.2 shows the liquid and vapor (molar) composition, as well as the reaction rate and temperature profiles along the reactive distillation column (Kiss 2011). The RD column is operated at ambient pressure, in the temperature range of 70–210 °C. As the reaction takes place mainly in the reactive zone, the reaction rate exhibits a maximum in the middle of the column. The concentration of water increases from the bottom to the top of the column, while the concentration of fatty ester increases from the top to bottom. Therefore, in the top of the column there is mainly water with negligible amounts of fatty acids, while in the bottom there is liquid fatty esters product with a very limited amount of methanol.

A rather similar RD process was proposed by Kiss (2010) for the production of biodiesel from hydrous bioethanol. Conventional integration of bioethanol and biodiesel plants employs the use of anhydrous ethanol in the biodiesel production

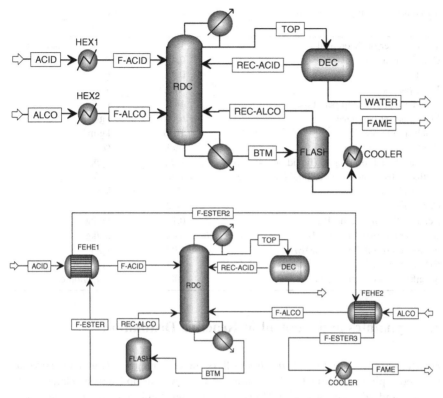

Fig. 5.1 Synthesis of fatty acid methyl esters (FAME) by reactive distillation: base case flowsheet (*top*), heat-integrated reactive distillation flowsheet (*btm*)

process—see Fig. 5.3 (Kiss 2010). However, the production of anhydrous bio-ethanol is very energy-demanding, especially due to the azeotropic distillation required to producing pure ethanol. The use of hydrous ethanol in the biodiesel production is preferable, but unfeasible in conventional processes due to the equilibrium limitations and the economic penalties caused by the additional process steps. To solve this problem, Kiss (2010) proposed a novel energy-efficient integrated production of biodiesel from hydrous bioethanol—as shown in Fig. 5.4. The key to success was a novel setup that combines the advantages of using solid catalysts with the integration of reaction and separation. This process eliminates all typical catalyst related operations, and efficiently uses the raw materials in a reactive separation unit that allows significant savings in the capital and operating costs. Rothenberg (2008) reported that a pilot plant based on FFA esterification with hydrous bioethanol in a RD column was built by Fertibom in Brazil.

Table 5.2 Design parameters for simulating the heat-integrated reactive distillation column

Parameter	Value	Units/remarks
Total number of theoretical stages	15	Reactive from 3 to 12
Column diameter	0.4	m
HETP	0.5	m
Valid phases	VLL	–
Volume liquid holdup per stage	18	L
Mass catalyst per stage	6.1	kg
Catalyst bulk density	1050	kg/m^3
Fatty acid conversion	>99.99	%
Fatty acid feed, on stage 3 (liquid, at 145 °C)	1167	kg/h
Methanol feed, on stage 10 (liquid, at 65.4 °C)	188	kg/h
Reboiler duty	136	kW
Condenser duty	−72	kW
Reflux ratio (mass ratio R/D)	0.10	kg/kg
Boil-up ratio (mass ratio V/B)	0.12	kg/kg
Production of biodiesel (FAME)	1250	kg/h
RD column productivity	20.4	kg FAME/kg cat/h
Specific energy requirements	108.8	kW h/ton ester

5.3 Dynamics and Control of Reactive Distillation

Heat-integrated reactive distillation offers indeed major advantages such as reduced capital investment and operating costs, as well as no catalyst-related waste streams and no soap formation. However, the controllability of the process is just as important as the savings in capital and operating costs. The constraint of stoichiometric ratio between reactants which guarantees product purity (Dimian et al. 2009; Bildea and Kiss 2011) must be fulfilled during the normal operation and the transitory regimes. In spite of the high degree of integration the heat-integrated RD process is well controllable—a key result being an efficient control structure that can ensure the reactants ratio required for the total conversion of fatty acids and for the prevention of difficult separations.

Figure 5.5 illustrates the plantwide control structure of the heat-integrated RD process proposed by Kiss (2011). The production rate is set by the flow rate of the Acid stream (controller FC1). The setpoint of the Alcohol feed flow controller FC2 is set, in a feed forward manner, by multiplying the Acid flow rate by the desired Alcohol:Acid ratio. It should be noted that setting the flow rates such that the stoichiometric ratio is fulfilled is not possible using only the flow controllers FC1 and FC2 because of unavoidable measurement or control implementation errors. For this reason, an additional concentration controller is necessary. An excess of Alcohol will have as result the complete consumption of Acid and a drop of the acid concentration at the bottom of the column. On the other hand, large quantities of fatty acid will be present when this reactant is in excess.

Table 5.3 Mass balance of a 10 ktpy biodiesel process based on integrated reactive-distillation

	F-ACID	F-ALCO	BTM	F-ESTER	REC-ACID	REC-ALCO	TOP	WATER	FAME
Temperature K	418.1	338.6	480.3	480.3	323.1	480.3	372.8	323.1	303.1
Pressure atm	1.036	1.036	1.017	0.85	1	0.85	0.987	1	0.85
Vapor Frac	0	0	0	0	0	1	0	0	0
Mass Flow kg/h	1168	188.7	1251.653	1250.428	0.108	1.225	105.154	105.047	1250.428
Volume Flow l/min	24.521	4.207	25.431	25.399	0.002	12.506	1.909	1.807	20.996
Enthalpy Gcal/h	−0.952	−0.329	−0.836	−0.835	0	−0.001	−0.39	−0.395	−0.957
Mass Flow kg/h									
Methanol	0	188.7	1.889	1.497	Trace	0.392	0.011	0.011	1.497
Acid	1168	0	0.144	0.144	0.096	0	0.11	0.015	0.144
Water	0	0	0.005	0.003	0.002	0.001	105.024	105.021	0.003
Ester-M	0	0	1249.616	1248.784	0.01	0.832	0.01	<0.001	1248.784
Mass Frac									
Methanol	0	1	0.002	0.001	17 ppm	0.32	100 ppm	100 ppm	0.001
Acid	1	0	115 ppm	115 ppm	0.889	60 ppm	0.001	0	115 ppm
Water	0	0	4 ppm	3 ppm	0.023	0.001	0.999	1	3 ppm
Ester-M	0	0	0.998	0.999	0.088	0.679	91 ppm	519 ppb	0.999
Mole Flow kmol/h									
Methanol	0	5.889	0.059	0.047	Trace	0.012	0	<0.001	0.047
Acid	5.831	0	0.001	0.001	<0.001	Trace	0.001	0	0.001
Water	0	0	<0.001	<0.001	<0.001	<0.001	5.83	5.83	<0.001
Ester-M	0	0	5.83	5.826	<0.001	0.004	<0.001	Trace	5.826
Mole Frac									
Methanol	0	1	0.01	0.008	89 ppm	0.755	56 ppm	56 ppm	0.008
Acid	1	0	122 ppm	122 ppm	0.725	23 ppm	95 ppm	13 ppm	122 ppm
Water	0	0	46 ppm	32 ppm	0.207	0.005	1	1	32 ppm
Ester-M	0	0	0.99	0.992	0.067	0.24	8 ppm	44 ppb	0.992

Fig. 5.2 Steady state
profiles: liquid and vapor
molar composition,
temperature and reaction rate
profiles along the reactive
distillation column

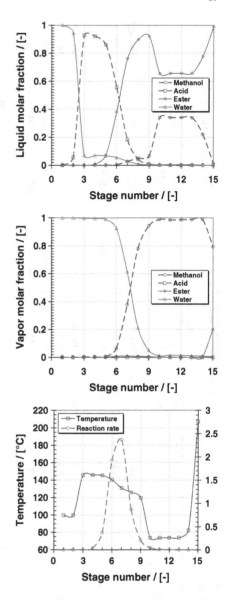

We conclude that the imbalance in the Alcohol:Acid ratio can be detected by measuring the concentration of Acid in the bottom outlet of the column (e.g. using a pH meter). Therefore, the concentration controller CC1 is used to give, in cascade manner, the Alcohol : Acid ratio. In this way, when production rate changes are implemented by changing the Acid flow rate, the correct ratio between reactant is achieved. The control of the distillation column is achieved by the pressure (PC1) and level (LC1, LC4) control loops. The reflux rate is kept constant because

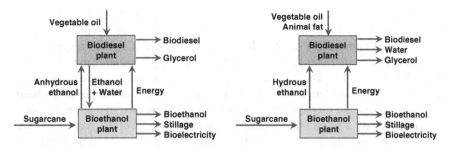

Fig. 5.3 Integrated bioethanol-biodiesel plant, using anhydrous (*left*) and hydrous ethanol (*right*)

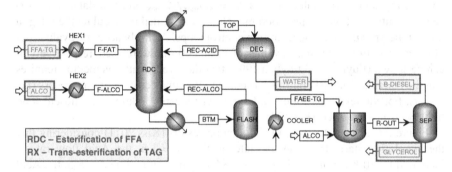

Fig. 5.4 Flowsheet of a two-step process for biodiesel production, able to convert any FFA level (0–100 %) feedstock to biodiesel

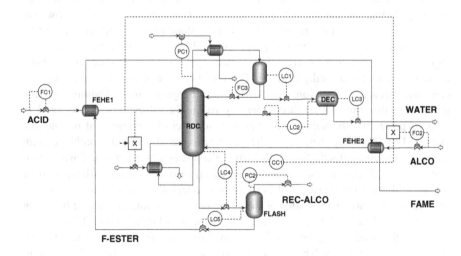

Fig. 5.5 Plantwide control structure of the heat-integrated RD flowsheet

of its small value, while the reboiler duty is ratioed to the Acid feed rate. The inventory of the two liquid phases in the decanter and the level and pressure in the flash are also controlled by standard loops (LC2, LC3 and LC5, PC2, respectively).

Details of the controller action and controller tuning parameters are presented in Table 5.4 (Kiss 2011). The flow control loops were assumed ideal due to their fast response. The control loops were tuned by a simple version of the direct synthesis method (Luyben and Luyben 1997). According to this method, the desired closed-loop response for a given input is specified. With the model of the process known, the required form and the tuning of the feedback controller are back-calculated. For all controllers, the acceptable control error, $\Delta\varepsilon_{max}$, and the maximum available control action, Δu_{max}, were specified. Then, the controller gain (expressed in engineering units) was calculated as $K_c = \Delta u_{max}/\Delta\varepsilon_{max}$ and translated into percentage units. First order open-loop models were assumed to calculate the integral time of the pressure and concentration control loops. As fairly accurate evaluations of the process time constants τ, values of 12 and 40 min were used, respectively. It can be shown (Luyben and Luyben 1997) that the direct synthesis method requires that the reset time of a PI controller is equal to the time constant of the process (i.e. $\tau_i = \tau$). For the level controllers, a large reset time $\tau_i = 60$min was chosen as no tight control is required (Kiss 2011).

Figure 5.6 depicts the dynamic simulation results (Kiss 2011) considering that the simulation starts from steady state: at time $t = 1$ h, the Acid flow rate is increased by 10 %, from 1168 kg/h to 1284.4 kg/h, then at time $t = 5$ h, the Acid flow rate is decreased to 1051.2 kg/h, representing a 10 % decrease with respect to the nominal value. The new production rate is achieved in about 2 h. The purity of FAME remains practically constant throughout the dynamic regime, the main impurity being methanol. Notably, the acid concentration stays below the 2000 ppm requirement of the ASTM D6751-08 standard (i.e. acid number <0.5).

Dual reactive distillation was proposed by Dimian et al. (2009) as a novel approach based on dual esterification of fatty acid with light and heavy alcohols, namely methanol and 2-ethylhexanol. These two complementary reactants have an equivalent reactive function but synergistic thermodynamic features. A key problem in the synthesis of fatty esters by RD is the effective water removal in view of protecting the solid catalyst and avoiding costly recovery of the alcohol excess. The RD setup behaves rather as a reactive absorption, combined with reactive azeotropic distillation with heavy alcohol as co-reactant and water-separation agent. Super acid solid catalyst based on sulfated zirconia, whose activity is comparable for the two alcohols, can be used at temperatures of 130–200 °C and moderated pressure. The control of the inventory of alcohols is realized by fixing the reflux of heavy alcohol and the light alcohol column inflow. This control strategy allows achieving both stoichiometric reactants feed rate and large flexibility in ester production. The distillation column for recovering light alcohol from water is no longer necessary. The result is a compact, efficient and easy-to-control multi-product reactive setup, as clearly illustrated by Fig. 5.7 (Dimian et al. 2009). The other design parameters of the RD column are conveniently provided in Table 5.5 (Dimian et al. 2009).

Table 5.4 Controller tuning parameters for the heat-integrated reactive distillation setup

Controller (direct action)	Range of manipulated variable	Range of controlled variable	Setpoint	Bias	Gain (%/ %)	Integral time (min)
Level LC1 (condenser drum)	0–210 kg/h	0–1 m	0.5 m	105 kg/h	1	60
Level LC2 (dec., org. lq.)	0–10 kg/h	0–1 m	0.7 m	1.3 kg/h	1	60
Level LC3(dec., aq. lq.)	0–210 kg/h	0–1 m	0.5 m	105 kg/h	1	60
Level LC4 (column sump)	0–250 kg/h	0–2 m	1 m	1251 kg/h	1	60
Level LC5 (flash)	0–2500 kg/h	0–2 m	1 m	1250 kg/h	1	60
Pressure PC1 (column)	0–12 kmol/h	0.9–1.1 bar	1 bar	6.25×10^4 kcal/ h	2	12
Pressure PC2 (flash)	0–0.1 kmol/h	0.7–1 bar	0.86 bar	0.026 mol/h	2	12
Mole fraction CC1 (acid, column btm)	0–200 ppm	0.05–0.25	100 ppm	0.1616	0.2	40

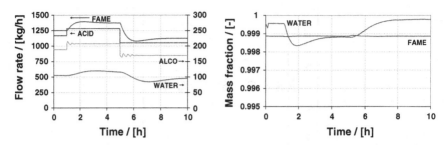

Fig. 5.6 Dynamic simulation results: flowrate and composition profiles at an acid flow rate disturbance of +10 % at 1 h, and −10 % at 5 h

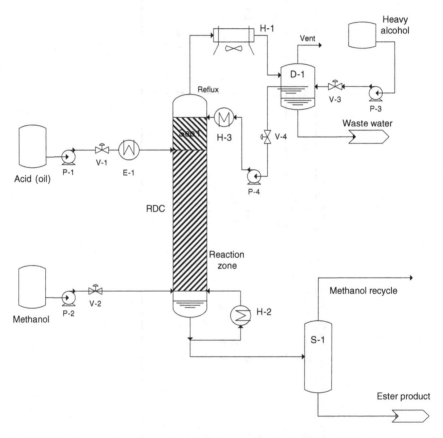

Fig. 5.7 Flowsheet configuration and control strategy of a dual reactive distillation setup

In the proposed control structure, the reactants are fed into the process in a ratio that satisfies the overall mass balance imposed by the reaction stoichiometry and the phase equilibrium at the top and the bottom of the RD column. In contrast, control structures fixing the feed rates of all reactants (acid, light- and heavy-alcohol) will

Table 5.5 Design parameters for simulating the dual reactive distillation column

Parameter	Value	Units/remarks
Number of theoretical stages	25	5–24 reactive
Lauric acid feed (on stage 5, at 3.5 bar, 150 °C)	100	kmol/h
Methanol feed (directly in reboiler, at 4 bar, 100 °C)	130	kmol/h
2-ethyl hexanol (fed in decanter, at 3.5 bar, 130 °C)	13.48	kmol/h
Catalyst bulk density	1050	kg/m^3
Volume holdup per stage	0.050	m^3
Mass catalyst per stage	55	kg
Reflux flow rate	2500	kg/h
Column diameter	1.2	m
HETP	0.5	m
Fatty acid conversion	>99.99	%
Reboiler duty	1750	kW
Condenser duty	−1492	kW
Production rate	22333	kg ester/h
Productivity of RD column	20.3	kg ester/kg cat/h
Bottom product composition (mass fraction)	650 ppb acid	kg/kg
	11 ppm water	kg/kg
	0.058 methanol	kg/kg
	0.788 methyl ester	kg/kg
	0.174 2EH ester	kg/kg
Specific energy requirements	166.8	kW h/ton ester

not work in the presence of small control implementation errors, the failure manifesting by the accumulation or depletion of one reactant (Kiss et al. 2007).

Figure 5.8 (top) compares the temperature profiles for the base case and for a 10 % increase of the lauric-acid flow rate with and without temperature control (Dimian et al. 2009). Accurate control of the lauric acid concentration in the bottom stream is achieved by using a concentration controller that prescribes the setpoint of the temperature controller in a cascade structure. For this control configuration, the change of the lauric acid feed flow rate leads to a change of the methyl-ester production rate. In contrast, when both flowrate ratios (i.e. lauric acid feed/methanol entering the column, and lauric acid feed/heavy alcohol reflux) are constant, the change of lauric acid feed flow rate leads to changes of both the methyl-ester and ethyl-hexyl ester production rates.

Figure 5.8 (middle) presents the performance of the control system for the following scenario: the simulation starts from the steady state (feed rate of lauric acid: 100 kmol/h) which is maintained for 0.5 h. Then, the feed rate of lauric acid is increased to 110 kmol/h and after 1 h is decreased to 90 kmol/h. Finally, the initial flow rate of 100 kmol/h is restored. The change of the acid feed flow rate leads to a change of the light ester production rate having the same magnitude of change, while the production rate of the heavy ester is constant. The dynamics is fast, only 20 min being necessary to achieve the new production rate. The amount

Fig. 5.8 Temperature
profiles along the RD column
(*top*). Dynamic simulation
results showing the increase
and decrease of the
production rate (*middle*).
Temperature and the
concentration of acid in the
bottom stream during
production rate changes
(*bottom*)

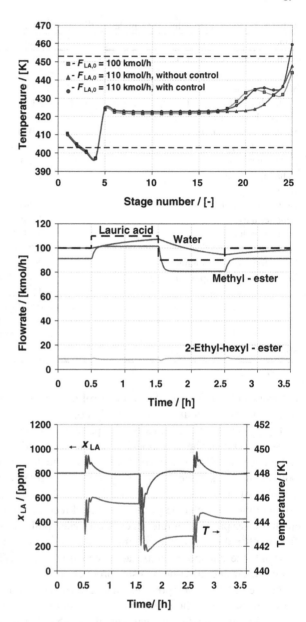

of water obtained at the top of the column reflects the amount of ester formed.
During the entire transient period, the concentration of water on the reactive trays
remains below the 2 wt% limit (Dimian et al. 2009).

Figure 5.8 (bottom) presents the temperature and the concentration of lauric
acid in the bottom stream, for the same scenario. Both variables remain very close
to the nominal values. Notably, the tuning of the controllers is not critical with

respect to the performance of the control system. In this case study, the parameters of the controllers were set as follows. The range of the controlled variable was set to the nominal value ± 10 °C for the temperature control loops, and the nominal value ± 50 % of it for the level control loops. For all loops, the range for the manipulated variable was set to twice the nominal value. The gain of the feedback controllers was set to 1 %/%. An integration time of 20 min was selected for the temperature controllers. In conclusion, the control structure proposed can achieve stable operation and it is able to modify the throughput, while keeping the characteristics of the products at their design values (Dimian et al. 2009).

At optimal operation the highest yield and purity can be achieved by using stoichiometric feeds in the desired ratio of fatty esters. At this point the amount of methanol lost in top is practically negligible. The heavy ester plays the role of a solvent and prevents escaping methanol in the top product. On the top stages the heavy alcohol enhances water concentration in the vapor phase, from which is separated by condensation and decanting, while heavy ester is produced in an amount proportional to the reflux flow rate. The optimal operation is based on controlling the inventory of reactants by using the principle of fixed recycle flows of co-reactants, in this case the reflux of the organic phase and the methanol inflow to the RDC. This strategy allows large changes in the production rate. The control strategy is generic and can be employed for esterification involving the formation of azeotropes, as for ethanol and (iso-)propanol. The overall result in integrating design and process control is a compact, efficient and easy-to-control multi-product reactive setup.

Note that in addition to the classic PID controllers, Model Predictive Control (MPC) is also a serious candidate for controlling non-linear processes such as reactive separations. The efficient implementations of MPC and dynamic optimization in industrial cases were already reported for several non-linear chemical processes including hybrid systems, such as reactive separations and in particular reactive distillation (Agachi et al. 2006; Nagy et al. 2007; Sharma and Singh 2010; Kiss 2010; Rewagad and Kiss 2012).

5.4 Complex Reactive Distillation Configurations

Distillation can be combined not only with reaction to form reactive distillation (RD), but also with distillation leading to a dividing-wall column (DWC) that is able to separate three high-purity streams within only one distillation tower (Olujic et al. 2003, 2009; van Diggelen et al. 2010; Kiss and Bildea 2011; Yildirim et al. 2011; Kiss and Ignat 2012). The double integration of reaction with distillation leads to a hybrid setup known as reactive dividing-wall column or R-DWC (Mueller and Kenig 2007; Kiss et al. 2009; Hernandez et al. 2009; Kiss and Suszwalak 2012; Kiss et al. 2012).

Hernandez et al. (2010) reported the esterification of lauric acid and methanol in a thermally coupled distillation sequence with a side rectifier and a Petlyuk setup, respectively. The former setup can produce ester with a high purity (around 99.9 %) and also pure water, the excess of methanol being recovered in the side rectifier. The results show that the energy requirement of the complex distillation sequence with a side rectifier can be reduced significantly by choosing the appropriate operating conditions. Moreover, dynamic tests for control of the composition of the ester and control of two temperatures indicate that it is possible to eliminate disturbances in the feed composition, while the composition of the biodiesel remains at the desired setpoint value.

Nguyen and Demirel (2011) studied the production of methyl dodecanoate (FAME) from lauric acid and methanol using sulfated zirconia as solid acid catalyst. Two distillation sequences were considered: (1) reactive distillation column followed by the methanol recovery column, and (2) thermally coupled reactive distillation and methanol recovery columns. Comparisons of the optimized sequences show that in the thermally coupled sequence, the energy requirements are reduced by 13.1 % in the reactive distillation column and 50 % in the methanol recovery column. The total energy losses for the columns are reduced by 281.35 kW corresponding to 21.7 % energy saving in the thermally coupled sequence. In addition, the composition profiles indicate that the thermally coupled RD column operates with a lower concentration of water in the reaction zone, which also reduces the catalytic deactivation.

Cossio-Vargas et al. (2011) explored the production of biodiesel from feedstock mixtures of fatty acids (oleic, linoleic, n-dodecanoic) using sulfuric acid as catalyst in reactive distillation sequences with thermal coupling. Chemical equilibrium was assumed on each reactive tray. The results indicate that the complex reactive distillation sequences can produce as bottoms product a mixture of fatty esters that can be used as biodiesel. In particular, the thermally coupled distillation sequence involving a side rectifier can handle the reaction and complete separation in accordance with process intensification principles.

In a follow-up study, Cossio-Vargas et al. (2012) investigated the reaction of fatty organic acids (a mixture similar to Jatropha curcas L. seed oil) with methanol to produce fatty esters in reactive thermally coupled distillation system. The results showed that it is indeed possible to obtain fatty esters as bottoms products with high purity suitable for use as biodiesel. Some additional benefits were found, for instance, recovering of the excess of methanol and removing of the water produced, achieving the total consumption of the acid. Also, energy savings of around 30 % were achieved in the case of thermally coupled reactive distillation systems, as compared to the classical two feeds reactive distillation column.

In a recent study, Kiss et al. (2012) proposed a novel biodiesel process based on a reactive dividing-wall column (DWC) that allows the use of only 15 % excess of methanol to convert completely the fatty acids feedstock (Fig. 5.9). FAME is produced as pure bottom product, water as side stream, while the methanol excess is recovered as distillate and recycled. The design employs a challenging global optimization problem with discrete and continuous decision variables. The optimal configuration was established by using simulated annealing as optimization method implemented in Mathworks Matlab, and coupled with rigorous simulations

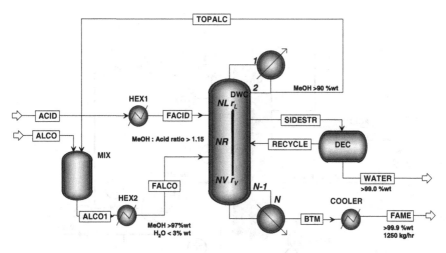

Fig. 5.9 Flowsheet and topology of a reactive DWC for FAME production: *1..N* number of stages, *NR* number of reactive stages, *NL* liquid split stage, *NV* vapor split stage

Table 5.6 Design parameters of an optimal reactive dividing-wall column for FAME synthesis by esterification (10 ktpy plant)

Design parameters	Value	Unit
Flowrate of ACID feed stream	1168.2	kg/h
Flowrate of ALCO feed stream	198.1	kg/h
Temperature of ACID feed stream	160	°C
Temperature of ALCO feed stream	69	°C
Pressure of feed stream	1.2	bar
Operating pressure	1	bar
Column diameter	1.1	m
Number of stages pre-fractionator side	12	–
Number of reactive stages pre-fractionator side	9	–
Total number of stages DWC	22	–
ACID feed stage pre-fractionator	4	–
ALCO feed stage pre-fractionator	9	–
Side stream withdrawal stage	16	–
Organic phase return stage	7	–
Wall position (from/to stage)	7–18	–
Liquid split ratio (r_L)	0.07	kg/kg
Vapor split ratio (r_V)	0.26	kg/kg
Methanol product purity	99.80/99.60	wt%/%mol
Water product purity	99.80/99.90	wt%/%mol
FAME product purity	99.99/99.99	wt%/%mol
HEX 1 duty	108	kW
HEX 2 duty	69	kW
Reboiler duty	212	kW
Condenser duty	−154	kW
Total heating duty	389	kW

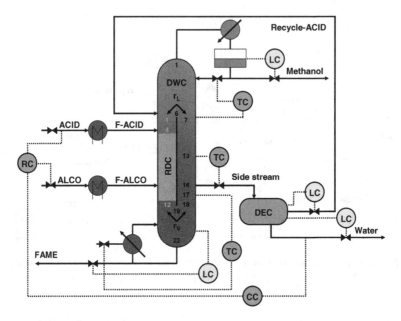

Fig. 5.10 Control structure of a reactive dividing-wall column for biodiesel production

carried out in AspenTech Aspen Plus. Along with the FAME production, the generated improved design alternatives allow lower investment costs and high energy savings.

The reactive DWC proposed by Kiss et al. (2012) was designed for a quaternary system—two products and two reactants. A critical aspect of the process is to ensure the full conversion of the FFA by having an excess of methanol and therefore, avoiding that unreacted FFA becomes an impurity in the bottom stream. However, the excess methanol can become an impurity in the side stream. To solve these problems, Ignat and Kiss (2013) investigated the optimal design and an efficient control strategy for a reactive DWC for FAME production. The initial results of the dynamic simulations indicated that the excess methanol becomes an impurity in the side stream under different scenarios. This problem arises from the fact that methanol is a light key component fed at the bottom of the column. An effective solution to overcome this problem is to feed the alcohol stream as vapor and to increase the acid inlet temperature. The reactive DWC design proposed by Kiss et al. (2012) was modified and optimized in terms of minimal energy use, by using the sequential quadratic programming (SQP) method coupled with the sensitivity analysis tool from Aspen Plus.

The main parameters of the optimal reactive DWC design are presented in Table 5.6 (Ignat and Kiss 2013). Compared to the previous study of Kiss et al. (2012), there are some key differences in the optimal R-DWC design that must be noted. In addition to the change of the feed streams thermal state, a key finding of Ignat and Kiss (2013) is that it is imperative to use a vapor feed of alcohol in order

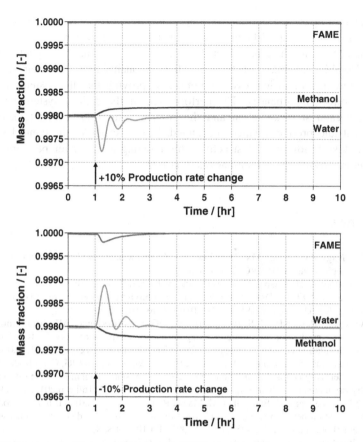

Fig. 5.11 Dynamic simulations results for a R-DWC, at production rate changes of ±10 %

to reach the product specifications. Moreover, while requiring 39 % less stages and 57 % less reactive stages, water and methanol are obtained as high purity products (99.8 %), with just a minor trade-off—only 1.5 % increase in the total heat duty.

Figure 5.10 illustrates the control structure proposed by Ignat and Kiss (2013). In practice, inferential temperature measurements are preferred to composition measurements. For the selection of the trays to control the temperature, several methods based on steady-state calculations are available—the slope criterion, invariant temperature criterion, sensitivity criterion and singular value decomposition method. In order to limit the amount of excess methanol that may pass below the partition wall of the column and therefore becomes an impurity in the side stream, an extra composition controller that measures the methanol mass fraction in the water stream is added to set the alcohol to acid ratio.

The level of the reflux drum, reboiler and the decanter can be controlled by the following manipulated variables D (distillate), B (bottoms), S (side stream) and Rec (recycle stream) respectively. The composition of the free product streams is controlled by the remaining variables: L (liquid reflux), S (side-stream) and

V (vapor boil-up). An extra control loop is needed in order to set the alcohol to acid ratio (*RxR*). Figure 5.11 presents the results of the dynamic simulations for industrially relevant disturbances such as production rate changes (Ignat and Kiss 2013). Notably, the mass fractions of all components return to their set point within short times and low overshooting, thus proving that the system can successfully reject disturbances. The performance of the control system was also tested for catalyst deactivation and the control system succeeded to keep the throughput and product quality unchanged. In this scenario, the control system compensates the lower catalyst activity by allowing a higher temperature inside the column—resulting in a similar reaction rate.

References

Agachi PS, Nagy ZK, Cristea MV, Imre-Lucaci A (2006) Model based control: case studies in process engineering. Wiley-VCH, Weinheim

Aspen technology (2010) Aspen Plus: user guide, vol 1 and 2. Mass, Burlington

Bildea CS, Kiss AA (2011) Dynamics and control of a biodiesel process by reactive absorption. Chem Eng Res Des 89:187–196. doi:10.1016/j.cherd.2010.05.007

Cossio-Vargas E, Hernandez S, Segovia-Hernandez JG, Cano-Rodriguez MI (2011) Simulation study of the production of biodiesel using feedstock mixtures of fatty acids in complex reactive distillation columns. Energy 36:6289–6297. doi:10.1016/j.energy.2011.10.005

Cossio-Vargas E, Barroso-Munoz FO, Hernandez S, Segovia-Hernandez JG, Cano-Rodriguez MI (2012) Thermally coupled distillation sequences: steady state simulation of the esterification of fatty organic acids. Chem Eng Process 62:176–182. doi:10.1016/j.cep.2012.08.004

da Silva ND, Santander CM, Batistella CM, Maciel R, Maciel MRW (2010) Biodiesel production from integration between reaction and separation system: reactive distillation process. Appl Biochem Biotechnol 161:245–254. doi:10.1007/s12010-009-8882-7

de Jong MC, Zondervan E, Dimian AC, de Haan AB (2010) Entrainer selection for the synthesis of fatty acid esters by entrainer-based reactive distillation. Chem Eng Res Des 88:34–44. doi:10.1021/ie100937p

Dimian AC, Bildea CS (2008) Chemical process design—computer-aided case studies. Wiley-VCH, Weinheim

Dimian AC, Bildea CS, Omota F, Kiss AA (2009) Innovative process for fatty acid esters by dual reactive distillation. Comput Chem Eng 33:743–750. doi:10.1016/j.compchemeng.2008.09.020

Dimian AC, Omota F, Bliek A (2004) Entrainer-enhanced reactive distillation. Chem Eng Process 43:411–420. doi:10.1016/S0255-2701(03)00125-9

Eleftheriades NM, von Blottnitz H (2013) Thermodynamic and kinetic considerations for biodiesel production by reactive distillation. Environ Prog Sustain Energy 32:373–376. doi:10.1002/ep.10621

Gomez-Castro FI, Rico-Ramirez V, Segovia-Hernandez JG, Hernandez S (2010) Feasibility study of a thermally coupled reactive distillation process for biodiesel production. Chem Eng Process 49:262–269. doi:10.1016/j.cep.2010.02.002

Gomez-Castro FI, Rico-Ramirez V, Segovia-Hernandez JG, Hernandez-Castro S (2011) Esterification of fatty acids in a thermally coupled reactive distillation column by the two-step supercritical methanol method. Chem Eng Res Des 89:480–490. doi:10.1016/j.cherd.2010.08.009

Hamed OA, Aly S, AbuKhousa E (1996) Heuristic approach for heat exchanger networks. Int J Energy Res 20:797–810. doi:10.1021/ie202171g

Harmsen GJ (2007) Reactive distillation: the front-runner of industrial process intensification—a full review of commercial applications, research, scale-up, design and operation. Chem Eng Process 46:774–780. doi:10.1016/j.cep.2007.06.005

He BB, Singh AP, Thompson JC (2005) Experimental optimization of a continuous-flow reactive distillation reactor for biodiesel production. Trans ASAE 48:2237–2243

He BB, Singh AP, Thompson JC (2006) A novel continuous-flow reactor using reactive distillation for biodiesel production. Trans ASABE 49:107–112

Hernandez S, Sandoval-Vergara R, Barroso-Munoz FO, Murrieta-Duenas R, Hernandez-Escoto H, Segovia-Hernandez JG, Rico-Ramirez V (2009) Reactive dividing wall distillation columns: simulation and implementation in a pilot plant. Chem Eng Process 48:250–258. doi:10.1016/j.cep.2008.03.015

Hernandez S, Segovia-Hernandez JG, Juarez-Trujillo L, Estrada-Pacheco EJ, Maya-Yescas R (2010) Design study of the control of a reactive thermally coupled distillation sequence for the esterification of fatty organic acids. Chem Eng Commun 198:1–18. doi: 10.1080/00986445.2010.493102

Ignat RM, Kiss AA (2013) Optimal design, dynamics and control of a reactive DWC for biodiesel production. Chem Eng Res Des 91:1760–1767. doi:10.1016/j.cherd.2013.02.009

Kiss AA (2010) Separative reactors for integrated production of bioethanol and biodiesel. Comput Chem Eng 34:812–820. doi:10.1016/j.compchemeng.2009.09.005

Kiss AA (2011) Heat-integrated reactive distillation process for synthesis of fatty esters. Fuel Process Technol 92:1288–1296. doi:10.1016/j.fuproc.2011.02.003

Kiss AA (2012) Applying reactive distillation. NPT Procestechnologie 19(1):22–24

Kiss AA (2013a) Advanced distillation technologies—design, control and applications. Wiley, UK

Kiss AA (2013b) Reactive distillation technology. In Boodhoo K, Harvey A (eds) Process intensification technologies for green chemistry: innovative engineering solutions for sustainable chemical processing. Wiley, New York, pp 251–274

Kiss AA (2013c) Novel applications of dividing-wall column technology to biofuel production processes. J Chem Technol Biotechnol 88:1387–1404. doi:10.1002/jctb.4108

Kiss AA, Bildea CS (2011) A control perspective on process intensification in dividing-wall columns. Chem Eng Process 50:281–292. doi:10.1016/j.cep.2011.01.011

Kiss AA, Bildea CS (2012) A review on biodiesel production by integrated reactive separation technologies. J Chem Technol Biotechnol 87:861–879. doi:10.1002/jctb.3785

Kiss AA, Ignat RM (2012) Enhanced methanol recovery and glycerol separation in biodiesel production—DWC makes it happen. Appl Energy 99:146–153. doi:10.1016/j.apenergy.2012.04.019

Kiss AA, Suszwalak DJ-PC (2012) Innovative dimethyl ether synthesis in a reactive dividing-wall column. Comput Chem Eng 38:74–81. doi:10.1016/j.compchemeng.2011.11.012

Kiss AA, Dimian AC, Rothenberg G (2006a) Solid acid catalysts for biodiesel production—towards sustainable energy. Adv Synth Catal 348:75–81. doi:10.1002/adsc.200505160

Kiss AA, Bildea CS, Dimian AC (2007) Design and control of recycle systems by non-linear analysis. Comput Chem Eng 31:601–611. doi:10.1016/j.compchemeng.2006.09.002

Kiss AA, Dimian AC, Rothenberg G (2008) Biodiesel by catalytic reactive distillation powered by metal oxides. Energy Fuels 22:598–604. doi:10.1021/ef700265y

Kiss AA, Omota F, Dimian AC, Rothenberg G (2006b) The heterogeneous advantage: biodiesel by catalytic reactive distillation. Top Catal 40:141–150. doi:10.1007/s11244-006-0116-4

Kiss AA, Pragt J, van Strien C (2009) Reactive dividing-wall columns—how to get more with less resources? Chem Eng Commun 196:1366–1374. doi:10.1080/00986440902935507

Kiss AA, Segovia-Hernandez JG, Bildea CS, Miranda-Galindo EY, Hernandez S (2012) Reactive DWC leading the way to FAME and fortune. Fuel 95:352–359. doi:10.1016/j.fuel.2011.12.064

Luyben WL, Luyben ML (1997) Essentials of process control. McGraw-Hill, New York, US

Luyben WL, Yu CC (2008) Reactive distillation design and control. Wiley-AIChE, Hoboken

Machado GD, Aranda DA, Castier M, Cabral VF, Cardozo L (2011) Computer simulation of fatty acid esterification in reactive distillation columns. Ind Eng Chem Res 50:10176–10184. doi:10.1021/ie102327y

Machado GD, Pessoa FLP, Castier M, Aranda DAG, Cabral VF, Cardozo-Filho L (2013) Biodiesel production by esterification of hydrolyzed soybean oil with ethanol in reactive distillation columns: simulation studies. Ind Eng Chem Res 52:9461–9469. doi:10.1021/ie400806q

Mueller I, Kenig EY (2007) Reactive distillation in a dividing wall column: rate-based modeling and simulation. Ind Eng Chem Res 46:3709–3719. doi:10.1021/ie0610344

Nagy ZK, Klein R, Kiss AA, Findeisen R (2007) Advanced control of a reactive distillation column. Comput Aided Chem Eng 24:805–810. doi:10.1016/S1570-7946(07)80157-X

Nguyen N, Demirel Y (2011) Using thermally coupled reactive distillation columns in biodiesel production. Energy 36:4838–4847. doi:10.1016/j.energy.2011.05.020

Noeres C, Kenig EY, Gorak A (2003) Modelling of reactive separation processes: reactive absorption and reactive distillation. Chem Eng Process 42:157–178. doi:10.1016/S0255-2701(02)00086-7

Noshadi I, Amin NAS, Parnas RS (2011) Continuous production of biodiesel from waste cooking oil in a reactive distillation column catalyzed by solid heteropolyacid: optimization using response surface methodology (RSM). Fuel 94:156–164. doi:10.1016/j.fuel.2011.10.018

Olujic Z, Jodecke M, Shilkin A, Schuch G, Kaibel B (2009) Equipment improvement trends in distillation. Chem Eng Process 48:1089–1104. doi:10.1016/j.cep.2009.03.004

Olujic Z, Kaibel B, Jansen H, Rietfort T, Zich E, Frey G (2003) Distillation column internals/configurations for process intensification. Chem Biochem Eng Q 17:301–309

Omota F, Dimian AC, Bliek A (2003a) Fatty acid esterification by reactive distillation. Part 1: equilibrium-based design. Chem Eng Sci 58:3159–3174. doi:10.1016/S0009-2509(03)00165-9

Omota F, Dimian AC, Bliek A (2003b) Fatty acid esterification by reactive distillation. Part 2: kinetics-based design for sulphated zirconia catalysts. Chem Eng Sci 58:3175–3185. doi:10.1016/S0009-2509(03)00154-4

Prasertsit K, Mueanmas C, Tongurai C (2013) Transesterification of palm oil with methanol in a reactive distillation column. Chem Eng Process 70:21–26. doi:10.1016/j.cep.2013.05.011

Rewagad RR, Kiss AA (2012) Dynamic optimization of a dividing-wall column using model predictive control. Chem Eng Sci 68:132–142. doi:10.1016/j.ces.2011.09.022

Rothenberg G (2008) Catalysis: concepts and green applications. Wiley-VCH, Weinheim

Schembecker G, Tlatlik S (2003) Process synthesis for reactive separations. Chem Eng Process 42:179–189. doi:10.1016/S0255-2701(02)00087-9

Sharma N, Singh K (2010) Control of reactive distillation column: a review. Int J Chem React Eng 8:R5

Simasatitkul L, Siricharnsakunchai P, Patcharavorachot Y, Assabumrungrat S, Arpornwichanop A (2011) Reactive distillation for biodiesel production from soybean oil. Korean J Chem Eng 28:649–655. doi:10.1007/s11814-010-0440-z

Steinigeweg S, Gmehling J (2003) Esterification of a fatty acid by reactive distillation. Ind Eng Chem Res 42:3612–3619. doi:10.1021/ie020925i

Sundmacher K, Kienle A (2003) Reactive distillation: status and future directions. Wiley-VCH, Weinheim, Germany

Suwannakarn K, Lotero E, Ngaosuwan K, Goodwin JG (2009) Simultaneous free fatty acid esterification and triglyceride transesterification using a solid acid catalyst with in situ removal of water and unreacted methanol. Ind Eng Chem Res 48:2810–2818. doi:10.1021/ie800889w

Thotla S, Mahajani S (2009) Reactive distillation with side draw. Chem Eng Process 48:927–937. doi:10.1016/j.cep.2008.12.007

van Diggelen RC, Kiss AA, Heemink AW (2010) Comparison of control strategies for dividing-wall columns. Ind Eng Chem Res 49:288–307. doi:10.1021/ie9010673

von Scala C, Moritz P, Fassler P (2003) Process for the continuous production of fatty acid esters via reactive distillation. Chimia 57:799–801. doi:10.2533/000942903777678461

Yildirim O, Kiss AA, Kenig EY (2011) Dividing wall columns in chemical process industry: a review on current activities. Sep Purif Technol 80:403–417. doi:10.1016/j.seppur.2011.05.009

Chapter 6
Reactive Absorption Technology

Abstract Reactive absorption (RA) is most commonly encountered for the separation and/or purification of gas mixtures, but RA is also applied in the production of bulk chemicals, such as nitric and sulfuric acid. Nowadays, reactive separations (RD and RA) using green catalysts offer great opportunities for manufacturing fatty esters. This chapter describes a novel RA process that is similar to RD, but is characterized by the absence of a condenser and reboiler, allowing high selectivity as no products are recycled in form of reflux or boil-up vapors, and no thermal degradation of the products occurs due to a low temperature profile in the column.

Reactive absorption (RA) is essentially a mature process that is known since the foundation of modern chemical industry. More recently, the role of RA as a core environmental protection process has grown up significantly, and nowadays reactive absorption is the most widely applied reactive separation process—see review paper of Yildirim et al. (2012).

As an industrial process, the most commonly encountered use of RA is for the separation and/or purification of a gas mixture by the absorption of part of the mixture (e.g., CO_2, H_2S, NOx and SOx) in a solvent that is regenerated afterwards. However, apart from gas cleaning, RA is also applied in the production of bulk chemicals, such as nitric and sulfuric acid. More recently, reactive separations (RD and RA) using green catalysts offer great opportunities for manufacturing fatty esters, involved in specialty chemicals and biodiesel production.

This chapter covers only simulation studies on using reactive absorption for producing biodiesel by fatty acids esterification, as no experimental data was reported so far. Kiss et al. (2008) and Dimian et al. (2009) noticed already that the reactive distillation configurations proposed by them employ extremely low reflux ratio, thus behaving rather as reactive absorption units instead of reactive distillation. Accordingly, Kiss (2009) proposed a reactive absorption setup that is characterized by the absence of a condenser and reboiler hence allowing high selectivity as no products are recycled in form of reflux or boil-up vapors, and no thermal degradation of the products occurs due to a low temperature profile in the column.

A. A. Kiss, *Process Intensification Technologies for Biodiesel Production,*
SpringerBriefs in Applied Sciences and Technology, DOI: 10.1007/978-3-319-03554-3_6,
© The Author(s) 2014

Kiss and Bildea (2011) proposed a novel energy-efficient reactive absorption (RA) process for biodiesel production, which is very well controllable in spite of the high degree of integration. The integration of reaction and separation into one unit combined with the use of a heterogeneous catalyst offers major advantages such as: reduced capital investment, low operating costs, simplified downstream processing steps as well as no catalyst-related waste streams and no soap formation. Rigorous process simulations in Aspen Plus were used to design and control a plant producing 10 ktpy biodiesel (equivalent to a rate of 1250 kg/h) by esterification of methanol with FFA, using sulfated zirconia as solid acid catalyst.

The conceptual design of the process is based on a reactive absorption column that integrates the reaction and separation steps into one operating unit. The chemical equilibrium is shifted towards products formation by continuous removal of the reaction products, instead of using an excess of a reactant—typically the alcohol. Figure 6.1 (left) presents the flowsheet of this process based on conventional reactive absorption, as reported by Kiss (2009). Fatty acid (as liquid) and methanol (as vapor) are fed at the top and bottom of the column, respectively. Because methanol is consumed on the reactive stages, the top product contains only water with some amounts of acid. This mixture is easily separated in a decanter, water being withdrawn as product while the acid is recycled. The bottom stream contains FAME with small amounts of methanol, which is separated in a simple flash and recycled. The reference flowsheet is relatively simple, with just a few operating units, two cold streams that need to be pre-heated (fatty acid and alcohol) and two hot streams that have to be cooled down (top water and bottom fatty esters). The heat-integration was carried out by applying previously reported heuristic rules (Hamed et al. 1996; Dimian and Bildea 2008). Consequently, two feed-effluent heat exchangers (FEHE) replace partially or totally each of the two heat exchangers HEX1 and HEX2. Figure 6.1 (right) illustrates the improved process including heat-integration around the RA column (Kiss and Bildea 2011). The hot liquid product of the FLASH, a mixture of fatty esters, is used to pre-heat and vaporize the alcohol feed stream. If production changes are expected, the nominal design should include a bypass of the hot stream (dashed line in Fig. 6.1, right) which can be used for control objectives.

The composition, temperature and rate profiles are plotted in Fig. 6.2 (Kiss and Bildea 2011). Table 6.1 lists the main design parameters, such as column size, catalyst loading, and feed condition, while Table 6.2 provides the complete mass balance (Kiss and Bildea 2011). High conversion of the reactants is achieved, with the productivity of the RA unit exceeding 19 kg fatty ester/kg catalyst/h. The purity specification is higher than 99.9 %wt for the final biodiesel product (FAME stream). The total amount of the optional recycle streams (REC-BTM) is not significant representing less than 0.9 % of the total biodiesel production rate.

Kiss and Bildea (2011) used sensitivity analysis to evaluate the range of the operating parameters: reactants ratios, temperature of feed streams, decanting temperature, flashing pressure, and recycle rates. Figure 6.3 shows that the optimal molar ratio of the reactants (alcohol:acid) is very close to the stoichiometric value of one (Kiss 2009; Kiss and Bildea 2011). In practice, using a very small excess of

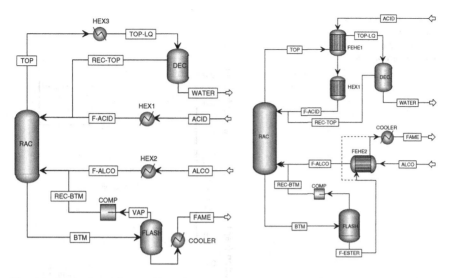

Fig. 6.1 Synthesis of fatty acid methyl esters (FAME) by reactive absorption: base case flowsheet (*left*) and heat-integrated process (*right*)

methanol (up to 1 %) or an efficient control structure is sufficient for the complete conversion of the free fatty acids.

Table 6.3 shows a head-to-head comparison of the heat-integrated RA process (Kiss and Bildea 2011) against the previously reported reference RD and RA processes (Kiss et al. 2008; Kiss 2009, 2011). These heating and cooling requirements are figures that ultimately translate into equipment size and cost. Remarkable, the energy demand is less than 22 kW h/ton biodiesel (34 kg steam/ton biodiesel). Also, compared to the reference RA base case (Kiss 2009), the heating and cooling requirements are significantly reduced by 85 and 90 %.

As heat-integrated RA offers major benefits, it comes as no surprise that several engineering companies from Asia (e.g. India, Indonesia, Malaysia) are interested in building this process.

However, the controllability of the process is just as important as the savings in capital and operating costs. In processes based on reactive distillation or absorption, feeding the reactants according to their stoichiometric ratio is essential to achieve high products purity (Dimian et al. 2009; Bildea and Kiss 2011). However, the integrated biodiesel processes based on RA have fewer degrees of freedom as compared to RD. This makes it very challenging to correctly set the reactants feed ratio and consequently avoiding the impurities in the products.

More recently, Kiss and Bildea (2011) proposed a rather simple but efficient plantwide control structure—illustrated here in Fig. 6.4—which can ensure the stoichiometric ratio of reactants and fulfills the excess of methanol operating constraint that is sufficient for the total conversion of the fatty acids. The production rate is set by the flow rate of the acid stream (controller FC1). After mixing with the recycle and passing the feed-effluent heat exchanger FEHE1, the

Fig. 6.2 Liquid-vapor
composition, temperature and
reaction rate profiles along
the reactive column

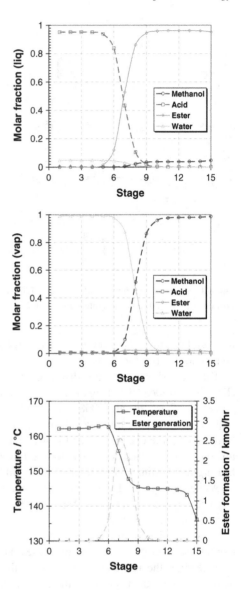

temperature of the acid fed to the column is controlled (by TC1) by manipulating
the duty of the heat exchanger HEX1. The flow rate of alcohol evaporated and fed
to the column is set by changing the split of the hot stream from the flash-outlet
stream (controller FC2). The temperature of the alcohol stream entering the column is therefore the boiling temperature. Because the hot FAME stream is not
available during the startup of the plant, the evaporator should include the option
of using another heat source. It should be noted that setting the flow rates such that
the stoichiometric ratio is fulfilled is not possible using only the flow controllers

Table 6.1 Design parameters for simulating the heat-integrated reactive absorption column

Parameter	Value	Units
Total number of theoretical stages	15	–
Number of reactive stages	10 (from 3 to 12)	–
Column diameter	0.4	m
HETP	0.6	m
Valid phases	VLL	–
Volume liquid holdup per stage	18	L
Mass catalyst per stage	6.5	kg
Catalyst bulk density	1050	kg/m^3
Fatty acid conversion	>99.99	%
Fatty acid feed (liquid, at 160 °C)	1167	kg/h
Methanol feed (vapor, at 65 °C)	188	kg/h
Production of biodiesel (FAME)	1250	kg/h
RA column productivity	19.2	kg FAME/kg cat/h

Table 6.2 Mass balance of a 10 ktpy FAME process based on integrated reactive-absorption

	F-ACID	F-ALCO	BTM	REC-BTM	REC-TOP	TOP	WATER	FAME
Temperature (°C)	160	65.4	136.2	146.2	51.8	162.1	51.8	30
Pressure (bar)	1.05	1.05	1.03	1.216	1	1	1	0.203
Vapor frac	0	1	0	1	0	1	0	0
Mole flow (kmol/h)	5.824	5.876	6.125	0.252	0.059	5.886	5.828	5.873
Mass flow (kg/h)	1166.7	188.3	1261.3	11.3	9.369	114.4	105.06	1250
Mass flow (kg/h)								
Methanol	0	188.3	9.125	7.544	0.002	0.103	0.101	1.581
Acid	116.74	0	Trace	Trace	9.218	9.233	0.016	Trace
Water	0	0	Trace	Trace	0.24	105.2	104.93	Trace
FAME	0	0	1252.2	3.764	0.846	0.846	Trace	1248.4
Mass fraction								
Methanol	0	1	0.007	0.667	172 ppm	894 ppm	965 ppm	0.001
Acid	1	0	Trace	Trace	0.894	0.08	148 ppm	Trace
Water	0	0	Trace	10 ppb	0.023	0.912	0.999	Trace
FAME	0	0	0.993	0.333	0.082	0.007	513 ppb	0.999
Mole fraction								
Methanol	0	1	0.046	0.931	873 ppm	546 ppm	0.001	0.008
Acid	1	0	Trace	Trace	0.726	0.008	13 ppm	Trace
Water	0	0	Trace	26 ppb	0.211	0.992	0.999	Trace
FAME	0	0	0.954	0.069	0.062	670 ppm	43 ppb	0.992

FC1 and FC2 because of unavoidable measurement or control implementation errors. For this reason, an additional concentration controller is necessary. An excess of alcohol will have as result the complete consumption of acid and a drop of the acid concentration at the bottom of the column. On the other hand, large

Fig. 6.3 Purity of *top* (water) and *bottom* (FAME) products: duty of heat exchangers (*top*), product and recycle flow rates (*middle*) versus molar reactants ratio alcohol:acid (*bottom*)

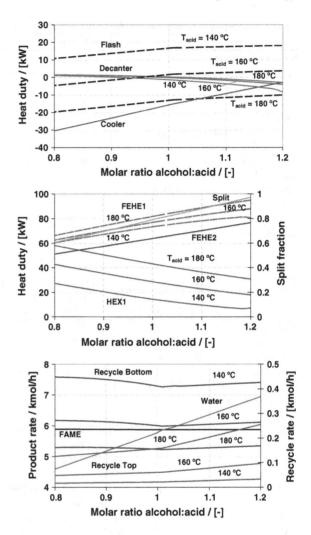

quantities of acid will be present when this reactant is in excess. We conclude that the imbalance in the alcohol:acid ratio can be detected by measuring the concentration of acid in the bottom outlet of the column. Therefore, the concentration controller CC2 is used to give, in a cascade manner, the correct setpoint to the alcohol flow rate controller FC2. In this way, when production rate changes are implemented by changing the acid flow rate, the correct ratio between reactants is achieved. However, for large production rate changes the overall reaction rate is not sufficiently large to guarantee the complete conversion of both reactants and high purity cannot be achieved. In particular, large amounts of methanol will be present in the top product of the absorption column. The concentration controller CC1 avoids this by increasing the setpoint of the temperature controller TC1.

Table 6.3 Comparison between (heat-integrated) reactive-absorption and reactive-distillation processes (at a production rate of 1250 kg/h fatty esters)

Equipment/parameter/units	RD	HI-RD	RA	HI-RA
Reactive column—reboiler duty (heater), kW	136	136	n/a	n/a
HEX-1 heat duty (fatty acid heater), kW	95	0	108	27
HEX-2 heat duty (methanol heater), kW	8	0	65	0
Reactive column—condenser duty (cooler), kW	−72	−72	n/a	n/a
HEX-3 water cooler/decanter, kW	−6	−6	−77	0
COOLER heat duty (biodiesel cooler), kW	−141	−38	−78	−14
FLASH heat duty (methanol recovery), kW	0	0	0	0
Compressor power (electricity), kW	0.6	0.6	0.6	0.6
Reactive column, number of reactive stages	10	10	10	10
Feed stage number, for acid/alcohol streams	3/10	3/10	1/15	1/15
Reactive column diameter, m	0.4	0.4	0.4	0.4
Reflux ratio (mass ratio R/D), kg/kg	0.10	0.10	n/a	n/a
Boil-up ratio (mass ratio V/B), kg/kg	0.12	0.12	n/a	n/a
Productivity, kg ester/kg catalyst/h	20.4	20.4	19.2	19.2
Energy requirements per ton biodiesel, kW h/ton FAME	191.2	108.8	138.4	21.6
Steam consumption, kg steam/ton FAME	295	168	214	34

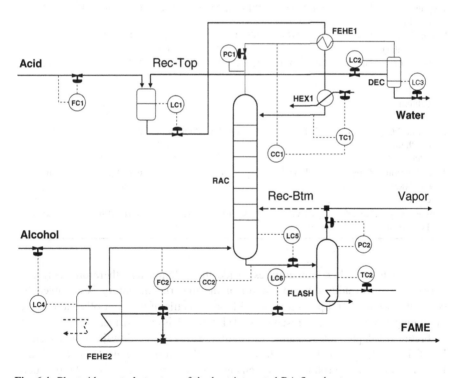

Fig. 6.4 Plantwide control structure of the heat-integrated RA flowsheet

Table 6.4 Controller tuning parameters for a heat-integrated reactive-absorption process

Controller	Control action	Range of manipulated variable	Range of controlled variable	Setpoint	Bias	Gain (%/%)	Integral time (min)
Level LC1 (buffer vessel)	Direct	0–2350 kg/h	0–1 m	0.5 m	1180 kg/h	1	–
Level LC2 (dec., org. lq.)	Direct	0–40 kg/h	0–1 m	0.5 m	14.2 kg/h	1	–
Level LC3 (dec., aq. lq.)	Direct	0–210 kg/h	0–1 m	0.45 m	105 kg/h	1	–
Level LC4 (alc. evap.)	Reverse	0–400 kg/h	0–2	1	197 kg/h	1	–
Level LC5 (col. sump)	Direct	0–2500 kg/h	0–1.75 m	0.875 m	1259 kg/h	10	60
Level LC6 (flash)	Direct	0–2490 kg/h	0–1 m	0.5 m	1245	1	–
Pressure PC1 (column)	Direct	0–12 kmol/h	0.95–1.05 bar	1 bar	5.91 kmol/h	1	12
Pressure PC2 (flash)	Reverse	0–1.22 kmol/h	0.18–0.22 bar	0.2 bar	0.29 kmol/h	1	12
Temp. TC1 (acid inlet)	Reverse	0–0.2 GJ/h	140–180 °C	Output of CC1	0.12 GJ/h	1	20
Mole fraction CC1 (methanol, column top)	Direct	140–180 °C	1–5×10^{-3}	3×10^{-3}	167.6 °C	0.1	40
Flow rate FC2 (methanol inlet)	Reverse	0.5–1	4–8 kmol/h	Output of CC2	0.9	1	20
Mole fraction CC2 (acid, col. btm.)	Direct	2–8 kmol/h	1–2×10^{-3}	1×10^{-3}	6.15 kmol/h	0.2	40
Temperature TC2 (flash)	Reverse	0–0.1 GJ/h	130–140 °C	137 °C	0.01 GJ/h	1	20

This leads to a higher temperatures inside the column, and therefore increased reaction rates. In addition, the control of the material inventory is achieved by conventional level and pressure control loops. Details of the controller action and the tuning parameters are presented in Table 6.4 (Kiss and Bildea 2011).

Figure 6.5 depicts the dynamic simulation results for the flowsheet presented in Fig. 6.1 (right), when the recycle of methanol vapors is considered (Kiss and Bildea 2011, 2013). Production rate changes are easily achieved and the products purity is maintained at high values, with the acid concentration in FAME below the 2000 ppm requirement of ASTM D6751-08 standard (i.e. acid number < 0.5).

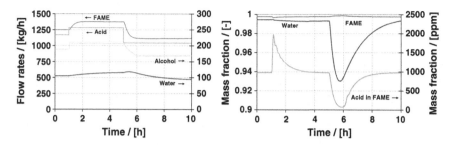

Fig. 6.5 Dynamic simulation results for the flowsheet with methanol recycle: acid flow rate disturbance of +10 % at 1 h, and −10 % at 5 h

Remarkable, the heat-integrated reactive absorption process described here eliminates all the conventional catalyst related operations, improves efficiency and considerably reduces the energy requirements for biodiesel production. Another important result is an efficient control structure that ensures the required reactants ratio, and fulfills the excess of methanol operating constraint that is sufficient for total conversion of fatty acids and for prevention of difficult separations. Moreover, in spite of the high degree of integration this reactive absorption process is very well controllable as illustrated by the results of rigorous dynamic simulations.

References

Bildea CS, Kiss AA (2011) Dynamics and control of a biodiesel process by reactive absorption. Chem Eng Res Des 89:187–196. doi:10.1016/j.cherd.2010.05.007

Dimian AC, Bildea CS (2008) Chemical process design—computer-aided case studies. Wiley-VCH, Weinheim

Dimian AC, Bildea CS, Omota F, Kiss AA (2009) Innovative process for fatty acid esters by dual reactive distillation. Comput Chem Eng 33:743–750. doi:10.1016/j.compchemeng.2008.09.020

Hamed OA, Aly S, AbuKhousa E (1996) Heuristic approach for heat exchanger networks. Int J Energy Res 20:797–810. doi:10.1021/ie202171g

Kiss AA (2009) Novel process for biodiesel by reactive absorption. Sep Purif Technol 69:280–287. doi:10.1016/j.seppur.2009.08.004

Kiss AA (2011) Heat-integrated reactive distillation process for synthesis of fatty esters. Fuel Process Technol 92:1288–1296. doi:10.1016/j.fuproc.2011.02.003

Kiss AA, Bildea CS (2011) Integrated reactive absorption process for synthesis of fatty esters. Bioresour Technol 102:490–498. doi:10.1016/j.biortech.2010.08.066

Kiss AA, Bildea CS (2013) Reactive absorption. In: Ramaswamy S, Huang H, Ramarao B (eds) Separation and purification technologies in biorefineries. Wiley, New York, pp 467–484

Kiss AA, Dimian AC, Rothenberg G (2008) Biodiesel by catalytic reactive distillation powered by metal oxides. Energy Fuels 22:598–604. doi:10.1021/ef700265y

Yildirim O, Kiss AA, Huser N, Lessmann K, Kenig EY (2012) Reactive absorption in chemical process industry: a review on current activities. Chem Eng J 213:371–391. doi:10.1016/j.cej.2012.09.121

Chapter 7
Reactive Extraction Technology

Abstract Reactive extraction is a process intensification alternative for conventional biodiesel processes, which effectively combines the extraction of oils using a solvent, with the (trans-) esterification reaction yielding fatty esters (biodiesel). This chapter gives an ample overview of using reactive extraction for biodiesel production—where typically the chemical reaction is combined with the separation of the reactants from the solid raw materials (plant seeds)—providing also a description of the experimental studies reported so far.

Reactive extraction is a process intensification alternative for conventional biodiesel processes, which effectively combines the extraction of oils using a solvent, with the (trans-) esterification reaction yielding fatty esters. Unlike reactive distillation or absorption—which combined reaction with the separation of products—in the case of reactive extraction for biodiesel production, the chemical reaction is typically combined with the separation of the reactants from the solid raw materials (seeds). Therefore, there is still need for an additional purification step to separate glycerol and yield pure fatty esters. Figure 7.1 illustrates the flow diagram of a generic reactive extraction process for biodiesel production.

This section describes only the experimental studies reported on producing fatty acid esters by trans-esterification of seed oils, as no process simulation studies were reported yet to the best of our knowledge. The main applications are the in situ extraction and trans-esterification of seed oils using acid/base homogeneous catalysts and various solvents (e.g. short chain esters).

Su et al. (2007) showed that by substituting short-chained alkyl acetates for short-chained alcohols as acyl acceptors for fatty acid esters production, the negative effects of glycerol and alcohol on lipase can be eliminated. Moreover, short-chained alkyl acetates, like other short-chained esters, are also suitable solvents for seed oil reactive extraction. In the work reported by Su et al. (2007), methyl acetate and ethyl acetate were used as extraction solvents and trans-esterification reagents for in situ reactive extraction of Pistacia chinensis Bunge seed and Jatropha curcas L seed. Fatty acid methyl esters (FAME) and ethyl esters (FAEE) were obtained with 5.3–22 % higher yields as compared to the extraction and subsequent trans-esterification process. Under optimal conditions, the highest

A. A. Kiss, *Process Intensification Technologies for Biodiesel Production*,
SpringerBriefs in Applied Sciences and Technology, DOI: 10.1007/978-3-319-03554-3_7,
© The Author(s) 2014

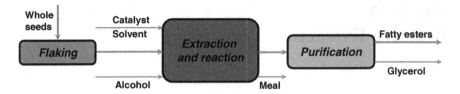

Fig. 7.1 Block flow diagram of a reactive extraction process for biodiesel production

yields for P. chinensis Bunge and J. curcas L methyl/ethyl esters were 92.8, 89.5, 86.1 and 87.2 %, respectively.

Su et al. (2009) proposed the use of dimethyl carbonate (DMC) or diethyl carbonate (DEC) as extraction solvent and trans-esterification reagent at the same time for in situ lipase-catalyzed reactive extraction of oilseeds for biodiesel production. Fatty acid methyl esters (FAME) and ethyl esters (FAEE) were respectively obtained with 15.7–31.7 % higher yields than those achieved by the regular two-step extraction/trans-esterification. The key parameters such as solvent/seed ratio and water content were also investigated to find their effects on the in situ reactive extraction (Su et al. 2009). Under optimized conditions, the highest yields were: 89.6 % for the Pistacia chinensis Bunge methyl ester, 90.7 % for the P. chinensis Bunge ethyl ester, 95.9 % for the Jatropha curcas L methyl ester and 94.5 % for the J. curcas L ethyl ester.

Kaul et al. (2010) explored reactive extraction (in situ) of Jatropha seeds in order to reduce the cost and increase the efficiency of biodiesel production. The oil from seeds was extracted and reacted in a single step. Experimental studies supported by high-performance liquid chromatography (HPLC) analysis were carried out to maximize the yield of biodiesel by varying the reaction parameters vs seed size (from <0.85 to >2.46 mm), seed/solvent ratio (from 1:2.6 to 1:7.8 w/w) and catalyst concentration (0.05–0.1 M). Conversion levels of 98 % were achieved under optimized conditions: seed size (>2.46 mm), seed/solvent ratio (1:7.8 w/w), catalyst concentration (0.1 M) and reaction time of 1 h.

Lim et al. (2010) proposed the use of supercritical reactive extraction from Jatropha curcas L. oil seeds as cost-effective processing technology for biodiesel production. Compared to traditional methods, the supercritical reactive extraction can successfully perform the extraction of oil and subsequent simultaneous esterification and trans-esterification process to FAME in relatively short times (45–80 min). Particle size of the seeds (0.5–2.0 mm) and the reaction temperature (200–300 °C) and pressure are the primary factors affecting the process. The optimal process conditions were reported to be: 300 °C reaction temperature, 240 MPa operating pressure, 10 ml/g methanol to solid ratio and 2.5 ml/g of n-hexane to seed ratio (Lim et al. 2010). The maximum oil extraction efficiency and FAME yield can reach up to 105.3 % v/v and 103.5 % w/w, respectively—which exceeded theoretical yield calculated based on n-hexane Soxhlet extraction of Jatropha oil seeds.

Shuit et al. (2010a, b) proposed, investigated and optimized a single step in situ reactive extraction of J. curcas L. seed to biodiesel. The size of the seed and reaction period affected significantly the yield of FAME and amount of oil extracted. The oil extraction efficiency and FAME yield reach 91.2 and 99.8 %, respectively, when using seed with size less than 0.355 mm and n-hexane as co-solvent with the following reaction conditions; reaction temperature of 60 °C, reaction period of 24 h, methanol to seed ratio of 7.5 ml/g and 15 %wt of H_2SO_4.

Moreover, design of experiments was used to study the effect of various process parameters on the yield of FAME. The process parameters studied include the reaction temperature (30–60 °C), methanol to seed ratio (5–20 mL/g), catalyst loading (5–30 wt%), and reaction time (1–24 h). The optimum reaction condition was then obtained by using response surface methodology coupled with central composite design. The reported results showed that an optimum biodiesel yield of 98.1 % can be obtained under the following reaction conditions: reaction temperature of 60 °C, methanol to seed ratio of 10.5 mL/g, 21.8 wt% of H_2SO_4, and reaction period of 10 h (Shuit et al. 2010a, b).

Dussan et al. (2010) prepared magnetic nanoparticles by co-precipitating Fe^{2+} and Fe^{3+} ions in a sodium hydroxide solution and used them as support for lipase. The lipase-coated particles were applied in a reactive extraction process that allowed the separation of products formed during trans-esterification. Kinetics data for triolein and ethanol consumption during the synthesis of ethyl oleate, as well as a thermodynamic phase equilibrium model (liquid–liquid) were used for the simulation of batch and continuous processes. The analysis proved the possibility of applying this biocatalytic system in the reactive zone using external magnetic fields. This approach implies new advantages in efficient location and use of lipases in column reactors for producing biodiesel. It is expected that reactive extraction can directly produce 77 % ethyl oleate (biodiesel) using lower ethanol/triolein ratios compared to 35–40 % purity in a normal stirred reactor with a higher ethanol/triolein ratio (Dussan et al. 2010).

Kasim and Harvey (2011) investigated the influence of a variety of parameters—including seeds size, agitation speed, reaction temperature, reaction time, catalyst concentration and molar ratio of alcohol to oil—for the biodiesel production by reactive extraction (in situ trans-esterification) of Jatropha curcas L. seeds. These parameters were varied in the ranges of <0.5–4 mm seeds particle size, 200–300 rpm agitation speed, 30–60 °C reaction temperature, 10–60 min reaction time, 0.1–0.2 N NaOH concentration and 100–600 molar ratio methanol to oil. The yield was found to be independent of the intensity of mixing once it reached 300 rpm, while the reaction temperature did not exhibit any significant effect on the yield. It was also demonstrated that the alkaline reactive extraction was complete in 20–30 min. The NaOH concentration can affect the yield in both positive and negative way. Low values (0.05 N) results in low yield, but higher concentrations (0.2 N) lead to formation of emulsions due to a saponification side reaction, adversely affecting the yield. In this case, a NaOH concentration of 0.15 N produced the highest yield. It was also reported that the biodiesel yield reached a constant state when the methanol to oil ratio was 400. The optimal conditions were: less than <0.71 mm seeds particle size, 300 rpm mixing speed,

30 °C temperature, 30 min reaction time, 0.15 N NaOH concentration and methanol to oil ratio of 400 (Kasim and Harvey 2011).

Gu et al. (2011) explored in situ reactive extraction of oilseeds with acyl acceptors as a sustainable and economically attractive biotechnological process for biodiesel synthesis. The presence of lipase activity in germinating oilseeds was detected, with an increase to a maximum activity after 4 days of germination. At that time, the germinated seeds featured slightly lower oil content relative to the un-germinated seeds. As such, an environmentally friendly and low-cost in situ self-catalytic process for biodiesel production was developed. The biodiesel was prepared by reactive extraction of germinated Jatropha curcas L. seeds with methanol in the absence of any catalyst and using n-hexane as reactive extraction solvent. The highest FAMEs yield of 87.6 % was achieved under the following optimal conditions: 2.5 ml/g n-hexane to germinated seed ratio, 1.5:1 molar ratio of methanol to germinated seed oil (containing 90 % FFA), reaction temperature of 35 °C, 2.9 % water content of germinated seed and 8 h reaction time. Such simple reactive extraction process without additional catalyst may greatly reduce the processing steps and costs of biodiesel production (Gu et al. 2011).

Pradhan et al. (2012) reported the production of biodiesel by reactive extraction of castor seed and determined the relationship between various important parameters of the RE process to obtain a high yield of FAME. Response surface methodology (RSM) was used to statistically analyze and optimize the operating parameters of the process. A central composite design (CCD) was adopted to study effects of catalyst concentration, methanol to oil molar ratio, reaction temperature and mixing intensity on yield of FAME. The results of RSM analysis indicated that catalyst concentration and methanol to oil molar ratio have most significant effect on the yield of FAME, with a good fitness of a second-order model. The interaction terms of catalyst concentration with both methanol to oil molar ratio and reaction temperature exhibited a positive effect delivering (P < 0.0001). Based on the second-order model, the optimum condition for this reaction were: methanol to oil molar ratio of 225:1, catalyst concentration 1.0 %wt of oil, reaction temperature of 55 °C and mixing intensity of 350 rpm. The mathematical model, represented in form of a quadratic polynomial equation, adequately described the ranges of the experimental parameters and provided a statistically accurate prediction of the optimum yield of FAME.

Zakaria and Harvey (2012) carried out a study to characterize the reactive extraction of rapeseed with methanol. The effects of process parameters on the yield, conversion and reaction rate differed substantially from the conventional trans-esterification, due to the dependence on both extraction and reaction. The rate of ester formation was mainly affected by the catalyst concentration, temperature and particle size. However, the equilibrium yield largely depends on the solvent to oil molar ratio. A high yield of ester (>85 %) was achieved only at high solvent to oil molar ratios, exceeding 475:1 (Zakaria and Harvey 2012). Parametric studies and light microscope images of reactively extracted seed suggested that reactive extraction occurs by trans-esterification of the oil inside the seed, followed by the diffusion of the products into the bulk solvent.

Bollin and Viamajala (2012) investigated the in situ production of fatty acid methyl esters directly from lipid-containing soy flour using a Lewis acidic ionic liquid: 1-ethyl-3-methyl-imidazolium chloroaluminate [EMIM]Cl·2AlCl$_3$ (N = 0.67). The system also contained methanol and dichloromethane, added as a co-solvent to reduce the viscosity of the ionic liquid (IL). The chloroaluminate, methanol, and dichloromethane formed a homogeneous reaction medium that facilitated a relatively rapid rate of FAME formation. The requirements for the ionic liquid, methanol, and cosolvent were determined, along with reaction temperature and time needed for high FAME yields. The results showed that FAME yields of >90 % can be achieved in 4 h, at 110 °C, with solids concentrations as high as 20 % (w/v). Furthermore, carbohydrates associated with the post-reaction residues remained chemically unmodified, which would allow for additional sugar-based co-product generation.

Porwal et al. (2012) reported the FAME production from Pongamia seeds (commonly known as Karanja) by reactive-extraction, which involves contacting the ground seeds directly with alcohol and catalyst, without prior extraction of the vegetable oil. Reaction parameters such as seed size (from less than 1 mm to over 2 mm), seed to solvent ratio (wt/wt) of 1:2–1:4, temperature of 30–60 °C and rate of mixing (250–550 rpm) were investigated. The maximum conversion of 98.5 % was achieved at: seed size <1 mm, seed to solvent ratio 1:4 wt/wt, 550 rpm rate of mixing, at a temperature of 60 °C, for a duration of 1 h, with 0.1 M catalyst (KOH) concentration, meeting the international (ASTM) and national (BIS) specifications.

Lim and Lee (2013a, b, c) carried out a supercritical reactive extraction with methanol in a high-pressure batch reactor to produce FAME from Jatropha curcas L. seeds. Supercritical reactive extraction takes the RE process one step further by substituting the role of catalyst with supercritical conditions to achieve higher yield and shorter processing time. Material and process parameters including space loading, solvent to seed ratio, co-solvent (n-hexane) to seed ratio, reaction temperature, reaction time and mixing intensity were varied one at a time and optimized based on two responses i.e. extraction efficiency, M-extract and FAME yield, F-y. The optimum responses for supercritical reactive extraction obtained were 104.17 % w/w and 99.67 % w/w (relative to 100 % lipid extraction with n-hexane) for M-extract and F-y respectively under the following conditions: 54.0 ml/g space loading, 5.0 ml/g methanol to seeds ratio, temperature of 300 °C, pressure of 9.5 MPa, 30 min reaction time and without n-hexane as co-solvent or any agitation source. This proved that supercritical reactive extraction is rather promising as another alternative for biodiesel production.

In another experimental study, Lim and Lee (2013a, b, c) reported a supercritical extraction and trans-esterification (SET) process for biodiesel production from oil seeds in which the solid oil-bearing material is used as the primary reactant together with short-chain alcohol directly in supercritical condition. The SET process with methanol was carried out in a high-pressure batch reactor to produce fatty acid methyl esters (FAME) from Jatropha curcas L. seeds (15.0 g feed, 300 °C, 5.0 ml/g methanol to solid ratio and 30 min). Different types of co-solvents (e.g. pentane, heptane, toluene, tetrahydrofuran, nitrogen and carbon dioxide) with varying amounts were added to study their influences towards the extraction efficiency. It was found that pentane and CO_2 provided higher responses (E-y: 102.6 % and 107.0 %, F-y: 100.4 % and 102.3 %) at concentration of 1.0 ml/g and 50 bar respectively. Addition of pentane and CO_2 lowered the critical conditions of the reactant mixture and achieved near optimum product yield at lower temperature (280 °C) and lower methanol to solid ratio (4.0 ml/g). The addition of appropriate co-solvents increases the extraction rate (solid-liquid) and enhances methanol-oil inter-phase miscibility during the reaction phase. This proved that SET process can be rather promising as another alternative route for biodiesel production.

In a follow-up study, Lim and Lee (2013a, b, c) performed the optimization of supercritical reactive extraction directly from Jatropha seeds in a high pressure batch reactor using Response Surface Methodology (RSM) coupled with Central Composite Rotatable Design (CCRD). Four primary variables—methanol solvent to solid ratio (SSR), reaction temperature, time and CO_2 initial pressure—were investigated under the proposed constraints. They found that all variables had significant effects towards fatty acid methyl esters (FAME) yield. Moreover, three interaction effects between the variables also played a major role in influencing the final FAME yield. Optimum FAME yield (92.0 %wt) was achieved under these conditions: 5.9 SSR, 300 °C, 12.3 min and 20 bar CO_2. The obtained FAME product fulfilled the existing international standard requirements. The characterization analysis proved that the solid residue can be burned as solid fuel in the form of biochar while the liquid product can be separated as specialty chemicals or burned as bio-oil for energy production.

Sulaiman et al. (2013) explored the possibility to obtain biodiesel from solid coconut waste, by reactive extraction. Solid coconut waste is produced after coconut milk extraction and may still contain 17–24 wt% extractable oil content. This study investigated the effect of catalyst amount, KOH concentration (0.8–2.0 %wt), temperature (55–65 °C) and mixing intensity (500-900 rpm) in order to optimize the reactive extraction process. Based on the Response Surface Methodology (RSM), the optimum conditions were: 2.0 wt% of KOH catalyst, 700 rpm of mixing intensity and a reaction temperature of 62 °C, resulting in 88.5 % biodiesel yield.

Madankar et al. (2013) investigated the feasibility of a reactive extraction process and explored the optimum reaction conditions for the production of castor oil methyl ester (COME) from castor seed. Reactive extraction of castor seed was carried out at varied reaction conditions—such as oil to methanol molar ratio of 1:50–1:250, catalyst concentration 0.5–2 %, temperature of 35–65 °C and rate of mixing 200–800 rpm—and the effect of these parameters on reactive extraction were investigated. The optimum reaction conditions for reactive extraction were found to be: 1 % KOH (wt basis of oil), reaction temperature 65 °C, reaction time 3 h, methanol to oil molar ratio 250:1 and mixing speed of 600 rpm. The kinetics of the reaction was studied and the activation energy was reported to be 38.916 kJ/mol. The potential use of COME as bio lubricants is promising due to its high viscosity, low pour point, and good lubricity. An economic analysis and sensitivity study of reactive extraction was carried out, leading to an estimated price of about $3.71/gal COME.

Jairurob et al. (2013a, b) investigated a single-step in situ extraction and trans-esterification (reactive extraction) of palm oil to biodiesel, exploring the effects of operating parameters of after-stripping sterilized palm fruit (A-sSPF). Process parameters included catalyst loading (1–4 % w/v), reaction time (8–11 h) and molar ratio of methanol to oil (170:1–260:1). The results showed that a biodiesel yield of 96.13 % (percentage weight of biodiesel per weight of palm oil in A-sSPF) can be obtained under the following reaction conditions: 3 % w/v of KOH, reaction period of 10 h, methanol to oil molar ratio of 230:1, as a temperature of 60 °C. The biodiesel yield based on palm fresh fruit bunch (FFB) was found to be 268.5 g biodiesel per kg FFB by the single-step reactive extraction, which is about 32 % as compared to 203.7 g biodiesel per kg FFB using the conventional two-step extraction and trans-esterification.

In a similar study, Jairurob et al. (2013a, b) investigated a single-step in situ extraction and trans-esterification (reactive extraction) method for the conversion of crude palm oil (CPO) in palm fruit fiber from mesocarp to biodiesel. The reaction parameters studied include the catalyst loadings (1–4 % KOH w/v), reaction time (8–11 h) and mole ratios of methanol to oil (147:1–225:1). An optimum biodiesel yield of 97.25 % w/w was obtained under these reaction conditions: catalyst loading of 3.85 % w/v of KOH, reaction period of 9.6 h, methanol to oil mole ratio of 225:1, and a reaction temperature of 60 °C. The biodiesel yield based on palm fresh fruit bunch (FFB) was found to be 272 g biodiesel per kg FFB in case of the single-step reactive extraction method, which is 55 % higher than 175 g biodiesel per kg FFB obtained by the conventional two-step extraction followed by reaction method.

More recently, Jurado et al. (2013) focused on the rigorous simulation of a hybrid reactive liquid-liquid extraction (LLX)—so not the usual seed extraction—as a novel alternative process for the production of biodiesel. The alkaline transesterification of vegetable oils was considered using methyl oleate (triolein) as major oil component and methanol as short chain alcohol. The process involves two zones in the extractive column: a reactive-extraction zone and a liquid-liquid extraction zone—as shown in Fig. 7.2 (Jurado et al. 2013). The reactants are fed in countercurrent, thereby improving the separation and the formation of products, glycerin and methyl oleate. Part of the glycerol obtained from the reactive section is used as an extractive agent in the extractive section. The process was simulated in Aspen Plus, assuming pseudo-homogeneous kinetics of alkaline catalysts. Remarkable, high conversion can be achieved, without a methanol excess to displace the equilibrium. The liquid-liquid equilibrium is effectively used in synergy with reaction, obtaining a very efficient separation.

The rigorous simulation proves that reactive extraction achieves has the potential to save more raw materials than other processes (Jurado et al. 2013). The energy usage is greatly reduced to about 20 % of the energy required in classic processes, thus reducing operating costs and environmental impact. This process also eliminates the need to use a large excess of methanol to shift the reaction equilibrium to the right and produce methyl esters (FAME) as a main product. Moreover, as all stages are integrated into a single unit, the equipment costs are

Fig. 7.2 Diagram of a
reactive extraction column
for biodiesel production

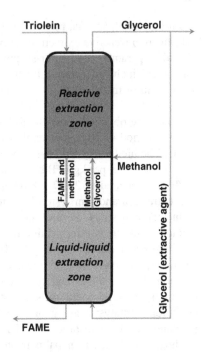

also drastically reduced. Glycerol and biodiesel (>99.8 %wt FAME) are obtained as high purity streams, therefore abiding to the EN14214 and ASTMD6751 standards. The high purity glycerol obtained in this process is useful for a high range of applications. Note that glycerol is currently a real waste problem in the biodiesel production, because its purity rarely exceeds 70 %wt (crude glycerol) hence not being usable as raw material in many industrial sectors.

In another study, Cadavid et al. (2013) reported a stepwise reaction-separation model for a counter-current reactive extraction column used in the homogeneous-base-catalyzed palm oil methanolysis. The parameters of the proposed model were identified from the results of a set of experimental tests, and a multi-criteria optimization of process conditions was performed using an evolutionary algorithm. The optimum conditions allow 97.7 % palm oil conversion and 99.5 % yield to FAME, as well as a high process productivity of 1.4 m^3 FAME / $h \cdot m^3$ in a single reaction step. As the counter-current reactive extraction process does not require the intermediate separation stages required in co-current and batch processes, the productivity is enhanced by a factor of 1.5 and 6.9 as compared to co-current and batch processes.

Based on the experimental reports described in the literature, it is clear that reactive extraction is mostly the preferred method of choice for the in situ extraction and trans-esterification of seed oils using acid or base homogeneous catalysts along with extraction solvents (e.g. short chain esters, DMC, DEC). Reactive extraction process is a promising alternative route for biodiesel production that not only reduces the number of processing steps and the capital investment but also increases the overall biodiesel yield.

References

Bollin PM, Viamajala S (2012) Reactive extraction of triglycerides as fatty acid methyl esters using Lewis acidic chloroaluminate ionic liquids. Energy Fuels 26:6411–6418. doi:10.1021/ef301101d

Cadavid JG, Godoy-Silva RD, Narvaez PC, Camargo M, Fonteix C (2013) Biodiesel production in a counter-current reactive extraction column: Modelling, parametric identification and optimization. Chem Eng J 228:717–723. doi: 10.1016/j.cej.2013.05.040

Dussan KJ, Cardona CA, Giraldo OH, Gutierrez LF, Perez VH (2010) Analysis of a reactive extraction process for biodiesel production using a lipase immobilized on magnetic nanostructures. Bioresour Technol 101:9542–9549. doi:10.1016/j.biortech.2010.07.044

Gu HQ, Jiang YJ, Zhou LY, Gao J (2011) Reactive extraction and in situ self-catalyzed methanolysis of germinated oilseed for biodiesel production. Energy Environ Sci 4:1337–1344. doi:10.1039/C0EE00350F

Jairurob P, Phalakornkule C, Na-udom A, Petiraksakul A (2013a) Reactive extraction of after-stripping sterilized palm fruit to biodiesel. Fuel 107:282–289. doi:10.1016/j.fuel.2013.01.051

Jairurob P, Phalakornkule C, Petiraksakul A (2013b) Single effects of reaction parameters in reactive extraction of palm fruit for biodiesel production. Chiang Mai J Sci 40:401–407

Jurado MBG, Plesu V, Ruiz JB, Ruiz AEB, Tuluc A, Llacuna JL (2013) Simulation of a hybrid reactive extraction unit: biodiesel synthesis. Chem Eng Trans 33:205–210. doi:10.3303/CET1335034

Kasim FH, Harvey AP (2011) Influence of various parameters on reactive extraction of *Jatropha curcas* L. for biodiesel production. Chem Eng J 171:1373–1378. doi:10.1016/j.cej.2011.05.050

Kaul S, Porwal J, Garg MO (2010) Parametric study of Jatropha seeds for biodiesel production by reactive extraction. J Am Oil Chem Soc 87:903–908. doi:10.1007/s11746-010-1566-1

Lim S, Lee KT (2013a) Process intensification for biodiesel production from *Jatropha curcas* L. seeds: Supercritical reactive extraction process parameters study. Appl Energy 103:712–720. doi:10.1016/j.apenergy.2012.11.024

Lim S, Lee KT (2013b) Influences of different co-solvents in simultaneous supercritical extraction and trans-esterification of *Jatropha curcas* L. seeds for the production of biodiesel. Chem Eng J 221:436–445. doi:10.1016/j.cej.2013.02.014

Lim S, Lee KT (2013c) Optimization of supercritical methanol reactive extraction by Response Surface Methodology and product characterization from *Jatropha curcas* L. seeds. Bioresour Technol 142:121–130. doi:10.1016/j.biortech.2013.05.010

Lim S, Hoong SS, Teong LK, Bhatia S (2010) Supercritical fluid reactive extraction of *Jatropha curcas* L. seeds with methanol: a novel biodiesel production method. Bioresour Technol 101:7169–7172. doi:10.1016/j.biortech.2010.03.134

Madankar CS, Pradhan S, Naik SN (2013) Parametric study of reactive extraction of castor seed (*Ricinus communis* L.) for methyl ester production and its potential use as bio lubricant. Ind Crops Prod 43:283–290. doi:10.1016/j.indcrop.2012.07.010

Porwal J, Bangwal D, Garg MO, Kaul S, Harvey AP, Lee JGM, Kasim FH, Eterigho EJ (2012) Reactive-extraction of pongamia seeds for biodiesel production. J Sci Ind Res 71:822–828

Pradhan S, Madankar CS, Mohanty P, Naik SN (2012) Optimization of reactive extraction of castor seed to produce biodiesel using response surface methodology. Fuel 97:848–855. doi:10.1016/j.fuel.2012.02.052

Shuit SH, Lee KT, Kamaruddin AH, Yusup S (2010a) Reactive extraction and in situ esterification of *Jatropha curcas* L. seeds for the production of biodiesel. Fuel 89:527–530. doi:10.1016/j.fuel.2009.07.011

Shuit SH, Lee KT, Kamaruddin AH, Yusup S (2010b) Reactive extraction of *Jatropha curcas* L. seed for production of biodiesel: process optimization study. Environ Sci Technol 44:4361–4367. doi:10.1021/es902608v

Su EZ, Xu WQ, Gao KL, Zheng Y, Wei DZ (2007) Lipase-catalyzed in situ reactive extraction of oilseeds with short-chained alkyl acetates for fatty acid esters production. J Mol Catal B-Enzym 48:28–32. doi:10.1016/j.molcatb.2007.06.003

Su EZ, You PY, Wei DZ (2009) In situ lipase-catalyzed reactive extraction of oilseeds with short-chained dialkyl carbonates for biodiesel production. Bioresour Technol 100:5813–5817. doi:10.1016/j.biortech.2009.06.077

Sulaiman S, Aziz ARA, Aroua MK (2013) Reactive extraction of solid coconut waste to produce biodiesel. J Taiwan Inst Chem Eng 44:233–238. doi:10.1016/j.jtice.2012.10.008

Zakaria R, Harvey AP (2012) Direct production of biodiesel from rapeseed by reactive extraction/in situ transesterification. Fuel Process Technol 102:53–60. doi:10.1016/j.fuproc.2012.04.026

Chapter 8
Membrane Reactors

Abstract Membrane reactors are an emerging technology for biodiesel production, combining both reaction and separation in a single unit aiming to overcome the equilibrium limitations by removal of reaction products. This chapter gives an overview on using membrane reactors in the production of biodiesel, where membrane plays an important role by removing glycerol from the product stream or by retaining the un-reacted triglycerides within the membrane. Pervaporation is also described as a process intensification technology applicable to biodiesel.

Membrane reactors combine the reaction and separation in a single unit with the goal of overcoming equilibrium limitations by removal of reaction products (Seidel-Morgenstern 2010; Water Environment Federation 2011; Oyama and Stagg-Williams 2011; Ismail and Matsuura 2012; Shuit et al. 2012). Membrane separation involves the use of a selective barrier (membrane) to regulate the transport of substances, such as gases, vapors and liquids, at different mass transfer rates that are controlled by the permeability of the barrier toward the feed components (Shuit et al. 2012). In the production of biodiesel, the membrane plays an important role by removing glycerol from the product stream or by retaining the un-reacted triglycerides within the membrane. Pervaporation is also applicable to biodiesel production.

Membrane reactors can be considered now an emerging technology for biodiesel production (Shuit et al. 2012). In order to successfully develop and commercialize membrane reactors in the biodiesel industry, knowledge is required in three major areas: catalysis, membrane technology and reactor engineering. Membrane technology attracted much interest for its ability to provide high purity and quality fuel besides its remarkable biodiesel yields. Membranes have many useful properties such as resistance to mechanical, chemical and thermal stress, high available surface area per unit volume, high selectivity, as well as the ability to control the components contact between the two phases. These properties make them potential candidates for both upstream and downstream biodiesel production and refining applications (Atadashi et al. 2011; Shuit et al. 2012). Additional key advantages are apparent in the biodiesel production, such as enhancement of the mass transfer between the immiscible methanol and oil phases and simplifying the

A. A. Kiss, *Process Intensification Technologies for Biodiesel Production*,
SpringerBriefs in Applied Sciences and Technology, DOI: 10.1007/978-3-319-03554-3_8,
© The Author(s) 2014

downstream processing operations. Atadashi et al. (2011) and Shuit et al. (2012) carried out comprehensive reviews of the main applications of membrane technology for the biodiesel production and refining technology.

The principle of membrane reactors for biodiesel production is shown in Fig. 8.1 (Kiss and Bildea 2012). The reaction occurs at the interface between oil and methanol phases. The reaction products (FAME and glycerol) and the catalyst are soluble in the methanol phase. Due to the smaller size of the molecules, the methanol/FAME/glycerol/catalyst phase selectively passes through the membrane into the permeate stream, while the oil droplets are retained due to the larger molecule size. Therefore, maintaining two liquid phases is a key factor for efficient functioning of the membrane reactors. Cheng et al. (2010) investigated the reaction conditions (temperature, methanol:oil ratio) required for a two-phase region reaction, and showed that increasing the residence time of the whole reactant system within the two-phase zone is beneficial for the separation operation through the membranes.

Dube et al. (2007) developed a 300 ml semi-batch, two-phase membrane reactor to produce FAME from canola oil and methanol. The carbon membrane had 6 mm internal diameter, 8 mm external diameter, 1200 mm length and 0.05 μm pore size. In each run, 100 ml of canola oil was used. Methanol and catalyst were fed at 2.5, 3.2 and 6.1 ml/min, while the catalyst concentration (sulfuric acid) was set to 0.5, 2, 4 and 6 %wt. The trans-esterification reaction was performed for 6 h at 60, 65 and 70 °C. The pressure difference between the reaction side and the permeation side was 138 kPa. Additional experiments were performed using 1 %wt NaOH as catalyst. Increases in temperature, catalyst concentration and feedstock (methanol/oil) flow rate significantly increased the conversion of oil to biodiesel. The novel reactor enabled the separation of reaction products (FAME/glycerol in methanol) from the original canola oil feed, yielding high purity biodiesel by shifting the reaction equilibrium. No sign of membrane degradation was observed after 10 months of reactor operation. The effect of membrane pore-size, methanol recycle and flux and residence time were investigated in three following papers (Cao et al. 2007, 2008a, b; Falahati and Tremblay 2012). Moreover, it was observed that the required concentration of the NaOH catalyst is 10-33 times lower than the one employed in the current industrial production of biodiesel (Tremblay et al. 2008).

Cao et al. (2008a, b) demonstrated that a membrane reactor can be continuously operated using a very broad range of feedstocks (soybean oil, canola oil, a hydrogenated palm oil/palm oil blend, yellow grease, brown grease,) at highly similar operating conditions to produce FAME. The overall internal volume of the reactor was 6.0 L. A Filtanium[TM] ceramic membrane constructed of a titanium oxide support and active layer was employed. The molecular weight cut-off (MWCO) of the membrane was 300 kDa. The oil and methanol/NaOH were fed at 45 and 38.5 g/min, respectively. The reaction took place at 65 °C. FAME produced using the membrane reactor easily met the free and total glycerol ASTM standards even for the low-grade lipid feedstock cases. In a follow-up paper (Cao et al. 2009), the same group of authors conducted the trans-esterification of canola

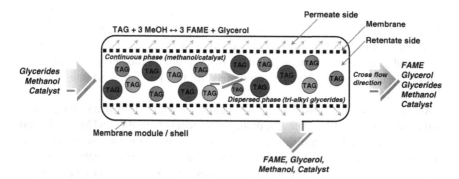

Fig. 8.1 Schematics of a membrane reactor for biodiesel production by trans-esterification of tri-alkyl glycerides

oil in the continuous membrane reactor, in the presence of NaOH as a catalyst with the goal of determining the forward and reverse rate constants of all three steps involved in the trans-esterification. The forward rate constants were greater than those previously reported for a batch process. This was attributed to the excellent mixing in the membrane reactor loop, the higher methanol/oil mole ratio used here, and the continuous removal of product from the reaction medium.

Machsun et al. (2010) used a biocatalytic membrane microreactor for continuous trans-esterification by utilizing an asymmetric membrane as an enzyme-carrier for immobilization. The performance of the biodiesel synthesis from triolein and methanol was studied. Trans-esterification was carried out by passing a solution of triolein and methanol through the membrane. The triolein conversion was about 80 % with a reaction time of 19 min. The system displayed good stability, with no activity decay over a period of 12 day with continuous operation. The results from the triolein trans-esterification clearly demonstrate the potential of an asymmetric membrane as an enzyme carrier material. Enzyme activity (mmol/h glipase) was approximately threefold higher than that of native free lipase.

Kapil et al. (2010) proposed to combine reaction and adsorption in simulated moving bed reactors (SMBR) for the production of high purity biodiesel. The study demonstrated that these processes have tremendous potential in terms of overcoming the low conversion and separation difficulties that are faced in conventional biodiesel production processes. Additionally, the process operating conditions were optimized to produce FAME at desired purity in a continuous mode. More than 90 % conversion of fatty acids and 80 % purity of FAME were achieved from an SMBR process operated at switching time of 900 s, length of 0.25 m, and feed, raffinate, and eluent flow rate ratios of 0.41, 0.49, and 0.75, for a given velocity of 2.4×10^{-4} m/s in the reaction zone.

Baroutian et al. (2011) developed a novel continuous reactor to produce high quality methyl esters from palm oil. A microporous TiO_2/Al_2O_3 membrane was packed with KOH catalyst supported on palm shell activated carbon. The effects of the reaction temperature, catalyst amount and cross flow circulation velocity on the production of biodiesel in the packed bed membrane reactor were investigated. The highest conversion of palm oil to biodiesel in the reactor was obtained at 70 °C employing 157.04 g catalyst per unit volume of reactor and 0.21 cm/s cross flow circulation velocity. The high quality of the palm oil biodiesel was proved by comparing the physico-chemical properties of the product with the standard specifications.

Atadashi et al. (2012) used a ceramic membrane with a pore size of 0.02 μm to purify crude biodiesel to achieve a product that meets both the ASTM D6751 and EN 14241 standards specifications. The process operating parameters such as trans-membrane pressure, flow rate and temperature were investigated. Application of central composite design (CCD) coupled with Response Surface Methodology (RSM) was found to provide clear understanding of the interaction between various process parameters. Then, the process operating parameters were optimized. The optimum conditions obtained were the following: trans-membrane pressure 2 bar, temperature 40 °C, and a flow rate of 150 L/min with corresponding permeate flux of 9.08 kg/m^{2}h. At these optimum conditions, the values of free glycerol (0.007 % wt) and potassium (0.297 mg/L) were all below the ASTM standard specifications for biodiesel fuel. In addition the physical properties of biodiesel at the optimum conditions met both standards.

Chong et al. (2013) conducted a thorough investigation on a membrane reactor for biodiesel production, considering the chemical phase equilibrium (CPE) via modeling analysis. A mathematical model was developed based on the modified Maxwell-Stefan model with the incorporation of CPE. The formation of triglycerides (TG) rich micelles dispersed in the continuous phase of MeOH was the most important hypothesis in the model development. The preliminary experimental results showed that the permeate compositions from the membrane reactor were closely related to the CPE, which was highly dependent on the MeOH to TG molar ratio. It was possible to obtain TG free permeate only if the continuous phase of MeOH was free from TG and the TG rich micelles were retained by the membrane. The model verification further confirmed the formation of micelles dispersed in the continuous MeOH phase within the feed side of the membrane reactor and the model was able to predict the performance of the membrane reactor for biodiesel production at an acceptable accuracy.

Shi et al. (2013) prepared a novel composite catalytic membrane (CCM), from sulfonated polyethersulfone (SPES) and polyethersulfone (PES) blend supported by non-woven fabrics, and used the CCM as a heterogeneous catalyst to produce biodiesel by esterification of oleic acid with methanol in a flow-through mode. A kinetic model of esterification was established based on a plug-flow assumption, and the effects of the CCM structure (thickness, area, porosity, etc.), the reaction temperature and the external and internal mass transfer resistances on esterification were investigated. The results showed that the CCM structure had a significant effect on the acid conversion. The external mass transfer resistance could be neglected when the flow rate was over 1.2 ml/min. The internal mass transfer resistance impacted on the conversion when membrane thickness was over 1.779 mm. The oleic acid conversion was kept over 98.0 % during 500 h of continuous operation. Moreover, the conversions obtained from the model were in good agreement with the experimental data.

Xu et al. (2013) reported the transesterification of soybean oil to produce biodiesel in a membrane reactor packed with shaped KF/Ca-Mg-Al hydrotalcite solid base as catalyst. The microfiltration ceramic membrane (length: 20 cm, inner/outer diameter: 6/10 mm) was used to retain the oil during the transesterification reaction. High quality biodiesel was produced in the fixed bed membrane reactor by coupling the heterogeneously alkali catalyzed reaction of transesterification with the separation process. Response surface methodology was employed to evaluate the effects of various factors, such as: reaction temperature, catalyst amount, and circulation velocity on biodiesel production. Under the optimum operating conditions (70 °C reaction temperature, 0.531 g/cm^3 catalyst amount, and 3.16 mL/min circulation velocity) it was possible to achieve a yield of 0.1820 g/min biodiesel, during 150 min circulation time.

More recently, Sajjadi et al. (2014) simulated the production of high quality fatty acid methyl ester (biodiesel) from palm oil in a micro porous ceramic membrane reactor. A TiO_2/Al_2O_3 ceramic membrane was used as separator and catalytic bed, and it was packed with potassium hydroxide catalyst supported on palm shell activated carbon. The investigation of component distribution within the system was not possible. Hence, computational fluid dynamics (CFD) analysis was used to predict the distribution of the fatty acid methyl ester and the other by-products in the membrane module. The Brinkman equation was used to simulate fluid flow within the porous media. In addition, the Maxwell-Stefan equation was applied for simulation of reaction kinetics and mass transfer. The combination of the mentioned models was solved mathematically by means of the finite element method and PARDISO algorithm. In addition, the effect of temperature on transesterification reaction has been examined. The CFD results indicated that increasing the reaction temperature leads to the same conversion in shorter time, or increase in temperature by 10 °C, results in 5 % growth of reaction for the same time period. The simulated liquid velocity within the system is in agreement with the experimental results, with 8.1 % deviation and 0.61 % overestimation in the reaction part.

Pervaporation was proposed as a clean technology to replace conventional energy intensive separation processes, such as evaporation and distillation. Separation by pervaporation does not rely on the relative volatilities but on the relative rates of permeation through a membrane. Pervaporation is also performed with a non-porous dense membrane that is usually made from a polymer or zeolite. It is typically applied to the dehydration of organic solvents, the removal of organic compounds from aqueous solutions and the separation of organic-organic mixtures. Based on the concept of the catalytic membrane, pervaporation can be an effective operation principle in biodiesel production (Shuit et al. 2012). In pervaporation processes—illustrated in Fig. 8.2 (Kiss and Bildea 2012)—the membrane acts as a barrier between the liquid retentate and the vapor permeate. The separation of components is based on the different rates of individual components transport through the membrane. For example, when hydrophilic membranes are used, water is selectively removed from the reaction mixture.

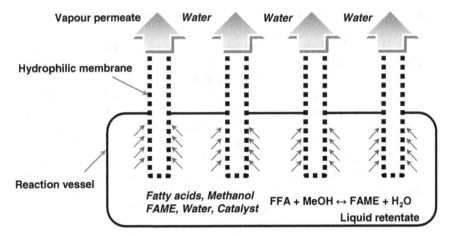

Fig. 8.2 Schematics of a membrane reactor for biodiesel production by esterification of fatty acids

Inoue et al. (2007) developed a pervaporation-aided dehydration reaction process that enabled ester condensation reactions stoichiometrically at nearly room temperature (lower than 323 K). The zeolite membranes consisted of acid-tolerant hydrophilic zeolites, with excellent permselectivity of water, even with coexistence of methanol. In the study of Sarkar et al. (2010), the authors used a polyvinyl alcohol (PVA) membrane cast on polyether sulfone (PES) to perform the esterification of oleic acid with methanol, catalyzed by sulfuric acid. A high conversion of 99.9 % was attained using 0.3 %wt H_2SO_4 at 65 °C, while the acid value of the methyl ester product was 0.2—which is lower than the 0.5 specification for biodiesel.

Figueiredo et al. (2010) investigated the pervaporation-assisted esterification of oleic acid and ethanol, catalyzed by Amberlyst 15. The hydrophilic polyvinyl alcohol (PVA) membrane was able to remove water from the reaction medium and consequently, the ester yield was increased. The potential of coupling the esterification and pervaporation was demonstrated, with a twofold increase in the reaction yield of ethyl oleate. Esterification of oleic acid with ethanol was also investigated under vapor permeation conditions (Okamoto et al. 1994), when the membrane is not in direct contact with the reaction liquid mixture and therefore it is expected to be more stable than in pervaporation. Almost complete conversion in a short time was observed, with a decrease of the initial molar ratio of ethanol to oleic acid.

Rewagad and Kiss (2012) proposed a generalized concept for modeling esterification using hetero/homogeneous catalysis, in a pervaporation membrane reactor (PVMR) that enhances the conversion by continuous removal of water byproduct. The system was described by the Maxwell-Stefan approach combined with the solution-diffusion model. The description of the membrane mass transport takes into account the interplay between adsorption and diffusion, as well as the

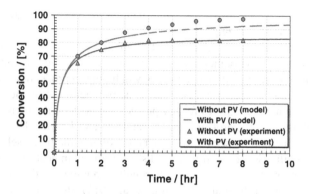

Fig. 8.3 Comparison of the model results against experimental data—conversion for the system with and without pervaporation (PV) membrane

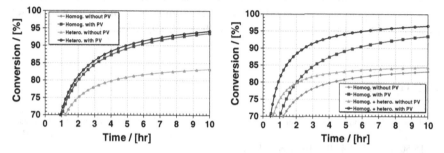

Fig. 8.4 Comparison of homogeneous and heterogeneous catalyzed systems, with and without pervaporation (PV) membrane—in terms of ethanol conversion

coupling between the diffusing species. This robust modeling approach was deployed in MatLab, where rigorous dynamic simulations of the PVMR were performed—providing a valuable insight into the operation, design and optimization of the process.

The system considered was the esterification reaction of oleic acid with ethanol. The comparison between various catalytic systems was supported by an appropriate sensitivity analysis study of the key design and operating parameters such as: permeability coefficient, membrane area-to-volume ratio, operating temperature and pressure. The simulations were also successfully validated against reported experimental data—see Fig. 8.3 (Rewagad and Kiss 2012)—thus proving the effectiveness of this modeling approach. The results showed that the pervaporation process can improve significantly the conversion of the fatty ester synthesis with 12.5, 13.2 and 16.1 % for the homogeneous, heterogeneous or combined catalytic system, respectively—as shown in Fig. 8.4 (Rewagad and Kiss 2012). Although the combined solid and liquid catalyst was the most beneficial option, the heterogeneous catalyst should be preferred from a practical point of view, due to benefits such as: no corrosion, no neutralization, no salt-waste streams and no additional separation steps required.

The main advantages of membrane technology concern the production of biodiesel in a more environmentally friendly and a cost-effective manner. The operation principles for membrane technology in biodiesel production are based on oil droplet size and the catalytic membrane. The production of biodiesel via catalytically inert membrane requires further purification because the permeate stream contains a mixture of glycerol, methanol, catalysts and FAME. Therefore, catalytically active membrane seems to be a better option for biodiesel production because less purification is required. Published findings indicate that membrane reactors have the potential to be a breakthrough technology in the biodiesel production industry (Atadashi et al. 2011; Shuit et al. 2012). However, the application of membrane technology to biodiesel production can be achieved only if the process parameters have been optimized. In addition to the typical parameters, such as reaction temperature, methanol-to-oil ratio and catalyst concentration, other process parameters such as reactant flow rate, trans-membrane pressure, membrane thickness and pore size, are reported to have great impacts on the yield of biodiesel as well, and they cannot be ignored (Atadashi et al. 2011; Shuit et al. 2012).

References

Atadashi IM, Aroua MK, Abdul Aziz AR, Sulaiman NMN (2011) Membrane biodiesel production and refining technology: a critical review. Renew Sustain Energy Rev 15:5051–5062. doi:10.1016/j.rser.2011.07.051

Atadashi IM, Aroua MK, Aziz ARA, Sulaiman NMN (2012) High quality biodiesel obtained through membrane technology. J Membr Sci 421:154–164. doi:10.1016/j.memsci.2012.07.006

Baroutian S, Aroua MK, Raman AAA, Sulaiman NMN (2011) A packed bed membrane reactor for production of biodiesel using activated carbon supported catalyst. Bioresour Technol 102:1095–1102. doi:10.1016/j.biortech.2010.08.076

Cao AP, Tremblay AY, Dube MA, Morse K (2007) Effect of membrane pore size on the performance of a membrane reactor for biodiesel production. Ind Eng Chem Res 46:52–58. doi:10.1021/ie060555o

Cao P, Dube MA, Tremblay AY (2008a) High-purity fatty acid methyl ester production from canola, soybean, palm, and yellow grease lipids by means of a membrane reactor. Biomass Bioenergy 32:1028–1036. doi:10.1016/j.biombioe.2008.01.020

Cao P, Dube MA, Tremblay AY (2008b) Methanol recycling in the production of biodiesel in a membrane reactor. Fuel 87:825–833. doi:10.1016/j.fuel.2007.05.048

Cao P, Dube MA, Tremblay AY (2009) Kinetics of canola oil transesterification in a membrane reactor. Ind Eng Chem Res 48:2533–2541. doi:10.1021/ie8009796

Cheng LH, Yen SY, Su LS, Chen J (2010) Study on membrane reactors for biodiesel production by phase behaviors of canola oil methanolysis in batch reactors. Bioresour Technol 101:6663–6668. doi:10.1016/j.biortech.2010.03.095

Chong MF, Chen JH, Oh PP, Chen ZS (2013) Modeling analysis of membrane reactor for biodiesel production. AIChE J 59:258–271. doi:10.1002/aic.13809

Dube MA, Tremblay AY, Liu J (2007) Biodiesel production using a membrane reactor. Bioresour Technol 98:639–647. doi:10.1016/j.biortech.2006.02.019

Falahati H, Tremblay AY (2012) The effect of flux and residence time in the production of biodiesel from various feedstocks using a membrane reactor. Fuel 91:126–133. doi:10.1016/j.fuel.2011.06.019

Figueiredo KCS, Salim VMM, Borges CP (2010) Ethyl oleate production by means of pervaporation-assisted esterification using heterogeneous catalysis. Braz J Chem Eng 27:609–617. doi:10.1590/S0104-66322010000400013

Inoue T, Nagase T, Hasegawa Y, Kiyozumi Y, Sato K, Nishioka M, Hamakawa S, Mizukami F (2007) Stoichiometric ester condensation reaction processes by pervaporative water removal via acid-tolerant zeolite membranes. Ind Eng Chem Res 46:3743–3750. doi:10.1021/ie0615178

Ismail AF, Matsuura T (2012) Sustainable membrane technology for energy, water, and environment. Wiley, New York

Kapil A, Bhat SA, Sadhukhan J (2010) Dynamic simulation of sorption enhanced reaction processes for biodiesel production. Ind Eng Chem Res 49:2326–2335. doi:10.1021/ie901225u

Kiss AA, Bildea CS (2012) A review on biodiesel production by integrated reactive separation technologies. J Chem Technol Biotechnol 87:861–879. doi:10.1002/jctb.3785

Machsun AL, Gozan M, Nasikin M, Setyahadi S, Yoo YJ (2010) Membrane microreactor in biocatalytic transesterification of triolein for biodiesel production. Biotechnol Bioprocess Eng 15:911–916. doi:10.1007/s12257-010-0151-7

Okamoto KI, Yamamoto M, Noda S, Semoto T, Otoshi Y, Tanaka K, Kita H (1994) Vapor-permeation-aided esterification of oleic acid. Ind Eng Chem Res 33:849–853. doi:10.1021/ie00028a010

Oyama ST, Stagg-Williams SM (2011) Inorganic polymeric and composite membranes: structure, function and other correlations. Elsevier, Amsterdam

Rewagad RR, Kiss AA (2012) Modeling and simulation of a pervaporation process for fatty ester synthesis. Chem Eng Commun 199:1357–1374. doi:10.1080/00986445.2012.660893

Sajjadi B, Abdul Raman AA, Baroutian S, Ibrahim S, Raja Ehsan Shah RSS (2014) 3D simulation of fatty acid methyl ester production in a packed membrane reactor. Fuel Process Technol 118:7–19. doi:10.1016/j.fuproc.2013.07.015

Sarkar B, Sridhar S, Saravanan K, Kale V (2010) Preparation of fatty acid methyl ester through temperature gradient driven pervaporation process. Chem Eng J 162:609–615. doi:10.1016/j.cej.2010.06.005

Seidel-Morgenstern A (2010) Membrane reactors: distributing reactants to improve selectivity and yield. Wiley-VCH, Germany

Shi W, He B, Cao Y, Li J, Yan F, Cui Z, Zou Z, Guo S, Qian X (2013) Continuous esterification to produce biodiesel by SPES/PES/NWF composite catalytic membrane in flow-through membrane reactor: experimental and kinetic studies. Bioresour Technol 129:100–107. doi:10.1016/j.biortech.2012.10.039

Shuit SH, Ong YT, Lee KT, Subhash B, Tan SH (2012) Membrane technology as a promising alternative in biodiesel production: a review. Biotechnol Adv 30:1364–1380. doi:10.1016/j.biotechadv.2012.02.009

Tremblay AY, Cao P, Dube MA (2008) Biodiesel production using ultralow catalyst concentrations. Energy Fuels 22:2748–2755. doi:10.1021/ef700769v

Water Environment Federation (2011) Membrane BioReactors. WEF Manual of Practice No. 36. McGraw-Hill Professional, US

Xu W, Gao LJ, Wang SC, Xiao GM (2013) Biodiesel production from soybean oil in a membrane reactor over hydrotalcite based catalyst: An optimization study. Energy Fuels 27:6738–6742. doi:10.1021/ef401823z

Chapter 9
Centrifugal Contact Separators

Abstract Centrifugal contact separator (CCS) is a process intensification approach integrating chemical reaction and centrifugal separation into a single apparatus. This chapter summarizes the recent applications to biodiesel production reported so far—mainly experimental studies based on the trans-esterification of virgin oil, using alkaline catalysts.

Centrifugal contact separator (CCS) is a process intensification approach that conveniently integrates chemical reaction and centrifugal separation into a single apparatus (Oh et al. 2012). The high gravity fields of centrifuges accelerate liquid-liquid separations by enhancing the specific gravity difference between the liquids involved. Two immiscible liquids with slight density differences can be rapidly and cleanly separated using a centrifugal contactor. Note that there are four basic operational functions of a centrifuge—with slight differences for each process: reactions, separations, extractions and water washes, multi-stage counter current extractions. Although CCS is one of the well-known examples of process intensification, only a few applications to biodiesel were reported so far—all of them being actually based on the trans-esterification of virgin oil, using alkaline catalysts.

Ondrey (2009) reported that Oak Ridge National Laboratory has developed a continuous process for producing biodiesel with a residence time of about 1 min, versus several hours for a conventional batch process. The process uses a centrifuge for simultaneous synthesis of biodiesel and separation of the biodiesel from the glycerol byproduct. The key to the process is a Couette reactor/separator, in which a food-grade soy oil and sodium methoxide (a mixture of methanol and sodium hydroxide) are fed separately and rapidly mixed at 3600 rpm. The reaction time is about 1 min, at which point the lighter methyl esters (biodiesel) and heavier glycerol are separated by the centrifugal action. The biodiesel yield for a 5:1 oil/methoxide ratio is about 95 %, with <1 % carryover of either phase into the other.

Kraai et al. (2008) proposed a new spin on catalysis, namely a table-top centrifugal contact separator that allows for fast continuous two-phase reactions to be performed by intimately mixing two immiscible phases and then separating them. Such a device has been used to produce biodiesel from sunflower oil and methanol

A. A. Kiss, *Process Intensification Technologies for Biodiesel Production*,
SpringerBriefs in Applied Sciences and Technology, DOI: 10.1007/978-3-319-03554-3_9,
© The Author(s) 2014

using NaOMe as catalyst, and also for the lipase-catalyzed esterification of oleic acid with nBuOH, both proceeding at high conversion.

McFarlane et al. (2010) and Kraai et al. (2009) reported the biodiesel synthesis by trans-esterification, using a centrifugal contact separator (CCS model CINC V02, available from CINC Industries, www.cincind.com) as illustrated in Fig. 9.1 (Abduh et al. 2013). The CCS used is a rotating centrifuge placed in a static reactor housing and operating in a once-through mode for both light and heavy liquid phases without recycle of the exit streams. Pure sunflower oil (preheated to 60 °C) was fed into the centrifugal reactor. The reaction was started by feeding into the CCS, the preheated solution (60 °C) of methanol (MeOH) and sodium methoxide (NaOMe), at a molar ratio of 6:1 MeOH:oil. Both immiscible liquids were dispersed in the annular zone between the static housing and the rotating centrifuge. The dispersion was then transferred into the hollow centrifuge, through a hole in the bottom plate, where phases were separated by centrifugal forces into light and heavy phases. After the reaction achieved the steady state, the glycerol and unreacted methanol flowed out of the heavy phase outlet, while the FAME with unreacted oil flowed out of the CCS through the light phase exit. Since a homogeneous catalyst was used, the dissolved catalyst was present in both outlet phases. Consequently, an additional refining or purification process is necessary (Oh et al. 2012).

As reported by Kraai et al. (2009), a steady state FAME yield of 96 % was achieved after about 30 min of reaction time in the CCS (CINC V02) operated at optimum conditions: rotational frequency 30 Hz, oil flow rate 12.6 mL/min, Na-OMe catalyst 1 w/w % of oil, reaction temperature of 75 °C. The yield productivity of CCS was at least comparable and likely higher than the conventional batch processes. Several methods were tested in order to enhance the FAME yield productivity, such as by increasing the reaction temperature, using higher catalyst dosage, adjusting oil feed flow rate and CCS rotational frequency. Due to the excessive MeOH evaporation at high temperature and solid soap formation in the centrifuge at higher intakes of catalyst, no yield enhancement was possible by increasing the reaction temperature or by using higher catalyst dosage. Moreover, the FAME yield drops when the oil feed flow rate increases due to the reduction of mean residence times of both light and heavy phases in the CCS, thus resulting in lower conversion. However, further reduction of feed flow rate is also not feasible due to the incomplete phase separation of both light and heavy outlet phases. On the other hand, the FAME yield increased with increasing the frequency up to a maximum of 30–40 Hz. A further increase of frequency leads to lower FAME yield. The increase of FAME yield is likely over the formation of small droplets in the annular zone and the high value of volumetric mass transfer coefficient. However, the FAME yield decreased at higher frequency values ($N > 40$ Hz). In the CCS, this biphasic trans-esterification reaction with relatively fast kinetics only takes place in a dispersion that is found in the annular zone as well as in some parts of the centrifuge. The volume of the dispersed phase in the annular zone is independent of frequency while the volume of the dispersed phase in the centrifuge is expected to be a function of frequency. According to Kraai et al. (2009), the settling velocity of the droplets at the dispersed zone of the centrifuge is

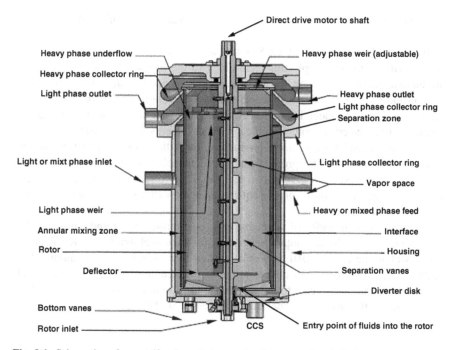

Fig. 9.1 Schematics of a centrifugal contact separator (*cross-sectional view*)

proportional to the difference in density, the angular momentum and the squared drop diameter. Therefore, due to the proportionality of the settling velocity with the angular momentum, it is expected that the dispersed phase volume in the centrifuge will be reduced considerably at high frequency (e.g. $N > 40$ Hz). As a result, by considering the strong reduction of the dispersed phase volume in the centrifuge, the FAME yield will be reduced.

Abduh et al. (2013) studied the synthesis of fatty acid ethyl esters (FAEE) from Jatropha curcas L. oil using sodium ethoxide as catalyst and ethanol as alcohol. Ethanol is accessible from biomass by fermentation being readily available in developing countries (e.g. Brazil) at a lower cost than methanol. The systems investigated were a batch reactor and a continuous centrifugal contactor separator. The effect of relevant process variables like rotational speed, temperature, catalyst concentration, and molar ratio of ethanol to oil was also investigated. The maximum yield of FAEE was 98 mol% for both the batch (70 °C, 600 rpm, 0.8 % w/w of sodium ethoxide) and CCS reactor configuration (60 °C, 2100 rpm, 1 % w/w of sodium ethoxide, oil feed 28 mL/min). The volumetric production rate of FAEE in the CCS at optimum conditions was 112 kg FAEE/m^3 liquid·min. The experimental volumetric productivity for the CCS was slightly lower than for the batch system. However, as the CCS is very compact, robust, and flexible in operation, it has high potential to be used in small scale mobile biodiesel units. Actually, such a mobile unit may consist of a cascade of two CCS in series, one for biodiesel synthesis and a second for a refining step with water/acid, followed by ethanol removal in a stripper.

Based on the review on the CCS (Oh et al. 2012), several concerns need to be considered. In order to apply this technology to the industrial scale for biodiesel production, modifications on the CCS are likely needed to achieve higher conversion and yield productivity. In the single-pass operation of CCS, the presence of the glycerol side product hinders a complete conversion due to the reversible nature of trans-esterification. Significant improvements in conversion and yield could be achieved by using multi-pass operation with two or three CCS units in series. In this manner, the product and unreacted oil from the first and second stages are separated from glycerol so that the forward reaction can be enhanced. The third stage can be used for subsequent aqueous wash to remove the remaining glycerol and catalyst.

So far, only high purity of vegetable oils (e.g. sunflower oil and soybean oils) were used in centrifugal contact separators for biodiesel synthesis (Kraai et al. 2009; McFarlane et al. 2010). It was suggested that the CCS can be modified to adapt various feedstock with a wide range of FFA content. The excess methanol into the CCS can be recycled to reduce the molar ratio of oil to methanol, and thus further reduce the overall production costs (Oh et al. 2012).

References

Abduh MY, van Ulden W, Kalpoe V, van de Bovenkamp HH, Manurung R, Heeres HJ (2013) Biodiesel synthesis from Jatropha curcas L. oil and ethanol in a continuous centrifugal contactor separator. Eur J Lipid Sci Technol 115:123–131. doi:10.1002/ejlt.201200173

Kraai GN, Van Zwol F, Schuur B, Heeres HJ, De Vries JG (2008) Two-phase (bio)catalytic reactions in a table-top centrifugal contact separator. Angew Chem Int Ed 47:3905–3908. doi:10.1002/anie.200705426

Kraai GN, Schuur B, van Zwol F, van de Bovenkamp HH, Heeres HJ (2009) Novel highly integrated biodiesel production technology in a centrifugal contactor separator device. Chem Eng J 154:384–389. doi:10.1016/j.cej.2009.04.047

McFarlane J, Tsouris C, Birdwell JF, Schuh DL, Jennings HL, Palmer Boitrago AM, Terpstra SM (2010) Production of biodiesel at the kinetic limit in a centrifugal reactor/separator. Ind Eng Chem Res 49:3160–3169. doi:10.1021/ie901229x

Oh PP, Lau HLN, Chen J, Chong MF, Choo YM (2012) A review on conventional technologies and emerging process intensification (PI) methods for biodiesel production. Renew Sustain Energy Rev 16:5131–5145. doi:10.1016/j.rser.2012.05.014

Ondrey G (2009) Chementator: this centrifugal reactor/separator speeds up biodiesel production. Chem Eng 116(2):1. (Brief article)

Chapter 10
Concluding Remarks

Biodiesel production can be tremendously enhanced by process intensification technologies. Each of the process intensification technologies has the potential to improve the production efficiency and therefore reduce the investment and operating costs of the process. Moreover, the enhancement in transport processes and higher reaction rates allow the change from conventional batch processes to continuous biodiesel production, thus ensuring large production rates (Qiu et al. 2010; Kiss and Bildea 2012).

Novel integrated processes for biodiesel production emerged lately, based on reactive separation technologies such as: reactive distillation, reactive absorption, reactive extraction, membrane reactors and centrifugal contact separators. The most preferred choice by now is reactive distillation that was successfully used for both esterification and trans-esterification reactions catalyzed by heterogeneous (e.g. ion exchange resins, sodium ethoxide, tungstated or sulfated zirconia, niobic oxide, mixed metal oxides) or homogeneous catalysts (e.g. NaOH, KOH)—as reported by numerous experimental studies, as well as process design and simulation studies (Kiss and Bildea 2012). However, reactive absorption was used so far only for the esterification of free fatty acids catalyzed by solid acids, while reactive extraction is clearly the preferred method of choice for the in situ extraction and trans-esterification of seed oils using acid or base homogeneous catalysts along with extraction solvents (e.g. short chain esters, DMC, DEC). Moreover, reactive liquid-liquid extraction (LLX) can be also used as PI alternative to conventional biodiesel processes based on trans-esterification.

Remarkable, the integration of reaction and separation into one unit overcomes all the equilibrium limitations and provides major benefits over conventional processes, especially when a solid (heterogeneous) catalyst is used and heat-integration is also employed:

- Elimination of all conventional catalyst-related operations such as: catalyst neutralization, separation and disposal of waste salts, waste water treatment.
- Efficient use of the raw materials and equipment: down to stoichiometric reactants ratio, complete conversion, low or no recycles, significantly high unit productivity.

A. A. Kiss, *Process Intensification Technologies for Biodiesel Production*, SpringerBriefs in Applied Sciences and Technology, DOI: 10.1007/978-3-319-03554-3_10, © The Author(s) 2014

- Multifunctional plants suitable for a large range of low quality feedstock with high FFA content, such as palm fatty acid distillate (PFAD), frying oils, animal tallow and waste vegetable oil. Improved product quality of the biodiesel fuel, meeting the ASTM standard.
- Drastically reduced capital and operating costs. Additional savings of 45–85 % are possible by using heat-integration around the reactive separation columns.

Membrane reactors are mainly used for trans-esterification processes, due to their excellent ability to restrict the passage of unreacted oils into the biodiesel product mixture thus providing a high-quality biodiesel fuel that meets the ASTM standard specification. In case of pervaporation processes based on fatty acids esterification, hydrophilic membranes are used to selectively remove water from the reaction mixture—the membrane acting as a barrier between the liquid retentate and the vapor permeate. Nevertheless, the development of inexpensive but very active heterogeneous catalysts is necessary in order to achieve high yields via membrane reactor production of biodiesel. Moreover, further research is needed to fully harness the unique characteristics of the membranes—high surface area per unit volume, high selectivity/conversion, capacity to control components contact, ability to recover important products, treat effluents (Atadashi et al. 2011; Shuit et al. 2012). In addition, critical attention must be paid to the fouling effects and stability of the proposed membranes.

Centrifugal contact separators (CCS) are also used in for biodiesel synthesis, but so far only with high purity vegetable oils (e.g. sunflower oil and soybean oils). However, CCS could be modified to adapt various feedstock sources with a wide range of FFA content. For example, the excess methanol into the CCS can be recycled to reduce the molar ratio of oil to methanol, and thus further reduce the overall production costs. Remarkable, the yield productivity of centrifugal contact separators was at least comparable and likely higher than the conventional batch processes. However, the use of a homogeneous catalyst leads to dissolved catalyst being present in both light and heavy outlet phases—this requiring an additional refining or purification process (Oh et al. 2012).

The process intensification technologies based on reactive separations clearly have the ability to improve the biodiesel production efficiency by integrating both reaction and separation into a single unit that allows the simultaneous production and removal of products, therefore enhancing the reaction rate, improving the productivity and selectivity, reducing the energy use, typically eliminating the need for solvents, intensifying the mass and heat transfer, and ultimately leading to extremely high-efficiency systems with 'green' engineering attributes.

References

Atadashi IM, Aroua MK, Abdul Aziz AR, Sulaiman NMN (2011) Membrane biodiesel production and refining technology: a critical review. Renew Sustain Energy Rev 15:5051–5062. doi:10.1016/j.rser.2011.07.051

Kiss AA, Bildea CS (2012) A review on biodiesel production by integrated reactive separation technologies. J Chem Technol Biotechnol 87:861–879. doi:10.1002/jctb.3785

Oh PP, Lau HLN, Chen J, Chong MF, Choo YM (2012) A review on conventional technologies and emerging process intensification (PI) methods for biodiesel production. Renew Sustain Energy Rev 16:5131–5145. doi:10.1016/j.rser.2012.05.014

Qiu ZY, Zhao LN, Weather L (2010) Process intensification technologies in continuous biodiesel production. Chem Eng Process 49:323–330. doi:10.1016/j.cep.2010.03.005

Shuit SH, Ong YT, Lee KT, Subhash B, Tan SH (2012) Membrane technology as a promising alternative in biodiesel production: a review. Biotechnol Adv 30:1364–1380. doi:10.1016/j.biotechadv.2012.02.009